NELSON MATHS

AUSTRALIAN CURRICULUM **NSW**

T0359232

Student Book

Pauline Rogers

NELSON
CENGAGE Learning·

Australia • Brazil • Japan • Korea • Mexico • Singapore • Spain • United Kingdom • United States

Contents

Identifying Place

1 Write the numeral shown on each abacus.

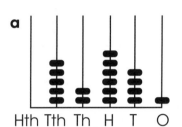
a
Hth Tth Th H T O

_____

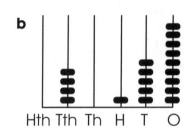
b
Hth Tth Th H T O

_____

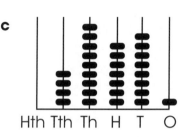
c
Hth Tth Th H T O

_____

2 State the value of the 6 in each number.

a 61 432 _____   **b** 11 056 _____

c 46 161 _____   **d** 16 528 _____

e 43 601 _____   **f** 50 116 _____

3 Circle the **largest** number in each pair.

a 29 006   2 998   **b** 42 119   41 911

c 16 832   17 852   **d** 60 479   60 794

e 185 609   185 690   **f** 4 006 142   4 600 142

4 Describe how you worked out which number was largest in Question 3c.

_____

5 Order the sets of numbers from **smallest** to **largest**.

a 16 487, 15 058, 17 598, 16 993   _____

_____

b 423 147, 406 102, 450 328, 424 199   _____

_____

c 307 421, 4 714 809, 3 074 456, 471 809   _____

_____

6 Shade the correct circle. To turn 9 million into 10 000 000 I would:

◯ add 100 000   ◯ subtract 1 000 000   ◯ add 1 000 000   ◯ subtract 100 000

Comparing Expanded Form

DATE:

1 Write the numeral for the numbers.

a 70 000 + 2 000 + 500 + 20 + 1   **b** 30 000 + 5 000 + 800 + 90 + 5

_____   _____

c 20 000 + 400 + 50 + 6   **d** 10 000 + 1 000 + 400 + 30 + 3

_____   _____

2 Write the numbers in expanded form.

a 60 419 _____

b 81 403 _____

3 Circle the **largest** number in each pair.

a 21 432   20 000 + 1 000 + 400 + 40 + 1

b 42 119   40 000 + 2 000 + 400 + 20 + 6

c 16 832   10 000 + 5 000 + 700 + 90 + 5

d 60 000 + 4 000 + 300 + 5   64 350

e 10 000 + 8 000 + 900 + 20 + 1   14 922

f 4 000 + 200 + 5   40 205

4 Describe how you worked out which number was **largest** in Question 3e.

_____

5 Order the sets of numbers from **smallest** to **largest**.

70 000 + 5 000 + 200 + 80 + 1

70 000 + 5 000 + 400 + 90 + 5

70 000 + 5 000 + 300 + 40 + 6

_____

Extension: Use all of the digits 4, 8, 7, 6, 5, 1 to:

a write the **smallest** number.   **b** write the **largest** number.

_____   _____

Unit **1**   **Place Value** (TRB pp. 20–23)
Whole numbers MA3-4NA orders, reads and represents integers of any size and describes properties of whole numbers

5

Which Animal?

Complete the gaps to find the unused number, and the animal that completes the joke.

a 10 658 = ten thousand, _____ hundred and fifty-eight

b 25 306 = twenty-five thousand, three hundred _____ six

c 61 259 = sixty _____ thousand, two hundred and fifty-nine

d 89 220 = eighty-nine thousand, two hundred and _____

e 44 244 = _____ -four thousand, two hundred and forty-four

f 16 225 = _____ thousand, two hundred and twenty-five

g 98 178 = ninety- _____ thousand, one hundred and seventy-eight

h 54 115 = fifty-four thousand, one hundred and _____

i 54_____24 = fifty-four thousand, nine hundred and twenty-four

j 11_____10 = eleven thousand, one hundred and ten

k 8456_____ = eighty-four thousand, five hundred and sixty

l 9_____325 = ninety-five thousand, three hundred and twenty-five

m _____4 235 = fourteen thousand, two hundred and thirty-five

What goes 99-clonk, 99-clonk, 99-clonk?

A _____ with a wooden leg!

DATE:

STUDENT ASSESSMENT

1 Write the numbers in the place-value chart.

   a 54 236   **b** 11 054   **c** 724 000

   d 4 782 368   **e** 921   **f** 31 270 356

	TM	M	HTh	TTh	Th	H	T	O
a								
b								
c								
d								
e								
f								

2 Write the numbers in expanded form.

   a 58 920 _____

   b 32 560 _____

3 Write the numbers in words.

   a 32 561 _____

   b 14 510 _____

4 Write the numeral for the numbers.

   a twenty-five thousand, three hundred and eleven _____

   b eighty-two thousand, one hundred and three _____

5 Order the following numbers from **smallest** to **largest**.

   a 56 123, 56 892, 56 741, 56 334

   _____

   b 2 114 021, 12 114 201, 12 114 210, 2 114 201

   _____

6 Circle the **larger** number in this pair. Describe how you worked it out.

   42 631   42 813 _____

   _____

Unit 1   **Place Value** (TRB pp. 20–23)
Whole numbers MA3-4NA orders, reads and respresents integers of any size and describes properties of whole numbers

7

Addition

1 Complete the equations.

a 475 + 321 =

b 436 + 122 =

c 1026 + 1421 =

d 3256 + 1003 =

e 4275 + 1012 =

f 4597 + 5301 =

2 Explain how you worked out the answer to Question 1e.

_____

3 Complete the equations.

a 419 + 47 =

b 328 + 453 =

c 459 + 448 =

d 103 + 298 =

4 Complete the equations.

a 5468
+ 3212

b 1003
+ 2463

c 3817
+ 619

d 4278
+ 1196

e 3471
+ 6385

f 14625
+ 13271

g 12685
+ 13976

h 43851
+ 11625

i 30095
+ 14721

j 46953
+ 42195

5 Describe how you worked out the answer to Question 4e.

_____

Extension: Write 4 different addition equations that equal 4263.

Unit **2** **Addition and Subtraction** (TRB pp. 24–27)
Addition and subtraction MA3-5NA selects and applies appropriate strategies for addition and subtraction with counting numbers of any size

Subtraction

1 Complete the equations.

a 498 – 321 = **b** 736 – 122 = **c** 1796 – 1421 =

d 3256 – 1013 = **e** 4275 – 2412 = **f** 9597 – 5301 =

2 Explain how you worked out the answer to Question 1e.

_____

3 Solve the subtraction problems.

a Find the difference between 459 and 867.

b Find 1359 minus 1170.

c Find 1798 less than 7689.

d What is 4569 take away 4308?

4 Complete the equations.

a	4710	**b**	6695	**c**	2786	**d**	4376	**e**	3721
	– 3248		– 1173		– 1597		– 2187		– 1845

f	84570	**g**	71719	**h**	39088	**i**	36273	**j**	58276
	– 32917		– 24462		– 26473		– 19485		– 39154

5 Explain how you worked out the answer to Question 4e.

_____

Extension: Use the numbers below to find a pair of numbers with:

a the **largest** difference. _____

b the **smallest** difference. _____

46538	79851	81689	47365	35801	68963

Unit 2 **Addition and Subtraction** (TRB pp. 24–27)
Addition and subtraction MA3-5NA selects and applies appropriate strategies for addition and subtraction with counting numbers of any size

9

Finding Addition and Subtraction

1 Use pairs of numbers from the table to complete the addition equations.

3 569	6 987	9 865	5 685	8 790	1 098	2 307	2 020	5 820	1 169	6 870	42
5 437	5 451	1 143	5 078	1 950	6 645	85	4 930	1 247	7 460	5 280	600

a

$$+ \underline{}$$
$$10\,763$$

b

$$+ \underline{}$$
$$8\,629$$

c

$$+ \underline{}$$
$$9\,006$$

d

$$+ \underline{}$$
$$12\,438$$

e

$$+ \underline{}$$
$$7\,743$$

f

$$+ \underline{}$$
$$6\,950$$

2 Use the remaining pairs from the table to complete the subtraction equations.

a e.g. 9 865

   – 1 143
   $\underline{}$

b

   – $\underline{}$

c

   – $\underline{}$

d

   – $\underline{}$

e

   – $\underline{}$

f

   – $\underline{}$

3 Fill in the missing digits to complete the equations.

a   2 4 7 9

    – 1 ☐ 4 6
    —————
    ☐ 7 ☐ ☐

b   2 ☐ 7 ☐

    – 1 6 ☐ 8
    —————
    ☐ 8 3 7

c   ☐ 4 ☐ 6

    – 2 ☐ 1 9
    —————
    1 7 5 ☐

Extension: Add and subtract numbers from the table to make 17 000. Record them on another sheet of paper.

Use a calculator to help you!

Unit **2**

STUDENT ASSESSMENT

1 Complete the equations.

a 458 + 368 =   **b** 782 + 119 =

c 1126 + 5935 =   **d** 1396 + 4876 =

e   24315   **f**   25678
   + 45917   + 65234
   _____   _____

2 Complete the equations.

a 598 − 325 =   **b** 781 − 498 =

c 1968 − 275 =   **d** 3756 − 1597 =

e   47951   **f**   78095
   − 22875   − 56017
   _____   _____

3 Explain how you worked out the answer to Question 2b. What strategy or strategies did you use?

_____

4 Solve the problems.

a A toy company needs 32568, 1095 and 15868 brochures to deliver to 3 different areas. How many brochures does it need to print? _____

b A stadium holds 85125 people. 92150 people want to attend a rock concert at the stadium. How many people could **not** go to the concert? _____

Unit **2**
Addition and Subtraction (TRB pp. 24–27)
Addition and subtraction MA3-5NA selects and applies appropriate strategies for addition and subtraction with counting numbers of any size

11

Multiples

1 Complete the number sequences.

   a 2, 4, 6, 8, _____, _____, _____, _____, _____

   b 3, 6, 9, 12, _____, _____, _____, _____, _____

   c 5, 10, 15, 20, _____, _____, _____, _____, _____

   d 6, 12, 18, 24, _____, _____, _____, _____, _____

   e 10, 20, 30, 40, _____, _____, _____, _____, _____

   f 9, 18, 27, 36, _____, _____, _____, _____, _____

2 Write the first 5 multiples of each number.

   a 5 _____

   b 6 _____

   c 4 _____

   d 7 _____

   e 11 _____

   f 20 _____

3 Write each number into the correct space in the table.

   50, 16, 8, 6, 18, 5, 33, 44, 15

	Multiples of 3	Multiples of 5	Multiples of 4
Less than 10			
Between 10 and 20			
Greater than 30			

4 Explain what **multiple** means. _____

_____

5 Is 112 a multiple of 4? Explain your answer.

_____

6 Are the multiples of 6 odd or even? Explain your answer.

_____

Factors

1 True or false (T or F)?

   a 6 is a factor of 24 _____   **b** 15 is a factor of 60 _____

   c 4 is a factor of 30 _____   **d** 7 is a factor of 28 _____

   e 3 is a factor of 90 _____   **f** 12 is a factor of 34 _____

2 List all the factors of each number.

   a 6 _____   **b** 18 _____

   c 30 _____   **d** 100 _____

   e 36 _____   **f** 45 _____

3 Explain how you found the answer to Question 2c.

_____

_____

4 Explain what **factor** means.

_____

_____

5 A **common factor** is a factor that is shared by more than 1 number.

For example: the factors of 6 are 1, 2, 3, 6

the factors of 8 are 1, 2, 4, 8

The common factors of 6 and 8 are 1 and 2.

Find the factors for each pair of numbers. Circle their **common factors**.

   a 12 _____   **b** 15 _____

      and   and

      36 _____   40 _____

   c 48 _____   **d** 20 _____

      and   and

      60 _____   28 _____

Unit **3** **Factors and Multiples** (TRB pp. 28–31)
Whole numbers MA3-4NA orders, reads and represents integers of any size and describes properties of whole numbers

13

Factors and Multiples

You may need: a calculator

1   Look at the table.

   a   Circle the factors of each of the given numbers.

   b   Shade the multiples of each of the given numbers.

Number					
12	3	36	60	10	6
20	40	5	8	4	100
8	2	4	16	32	81
15	30	3	45	10	5

2   Look at the numbers 77, 902, 278, 124, 125.

   a   Find the number that is divisible by 4. _____

   b   List all of that number's factors. _____

3   Draw 2 different factor trees for the number 20.

20   20

4   Write the first 10 multiples of each number.

   a   3 _____

   b   10 _____

   c   25 _____

   d   100 _____

DATE:

STUDENT ASSESSMENT

1 Write the first 5 multiples of each number.

    a 3 _____

    b 8 _____

    c 11 _____

    d 6 _____

    e 12 _____

2 List all the factors of each number.

    a 44 _____

    b 60 _____

    c 36 _____

    d 80 _____

3 List all the factors of 15 and 24. Circle the common factors.

    15 _____

    24 _____

4 Explain the difference between **factors** and **multiples**.

_____

_____

5 Draw a factor tree for 120.

120

Unit

3

Factors and Multiples (TRB pp. 28–31)
Whole numbers MA3-4NA orders, reads and respresents integers of any size and describes properties of whole numbers

15

Length

DATE:

You will need: a ruler

1 Tick the most appropriate unit of measurement for each item.

	Item	mm	cm	m	km
a	Length of a highway				
b	Thickness of a mobile phone				
c	Length of a classroom				
d	Length of a pencil				
e	Distance from Perth to Sydney				
f	Thickness of a DVD				

2 Write the lengths shown on the ruler.

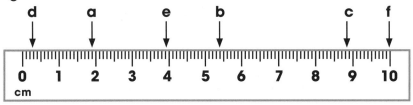

a _____ cm _____ mm **b** _____ cm _____ mm **c** _____ cm _____ mm

d _____ cm _____ mm **e** _____ cm _____ mm **f** _____ cm _____ mm

3 Measure each line to the nearest mm and cm.

a ▬▬▬▬▬▬▬▬▬

b ▬▬▬▬▬▬▬▬▬▬▬▬▬

c ▬▬▬▬

d ▬

e ▬▬▬▬▬▬▬

f ▬▬▬▬▬▬▬▬▬▬▬

4 Name items related to sport that could be measured in:

a millimetres _____ **b** centimetres _____

c metres _____ **d** kilometres _____

Extension: Measure the curled line to the nearest centimetre. _____

Length, Area and Volume (TRB pp. 32–35)
Length MA3-9MG selects and uses the appropriate unit and device to measure lengths and distances, calculates perimeters, and converts between units of length

Informal Area

You will need: a ruler

1 Find the area of each shape by counting the squares. Write the answer inside each shape.

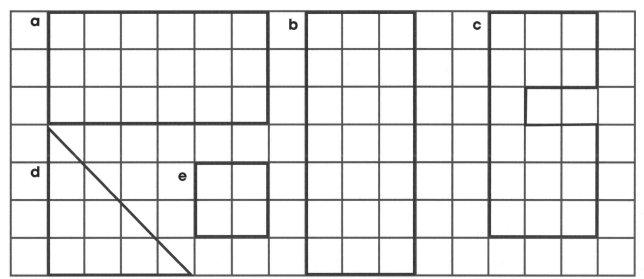

2 Find the area of each shape. Write the answer inside each shape.

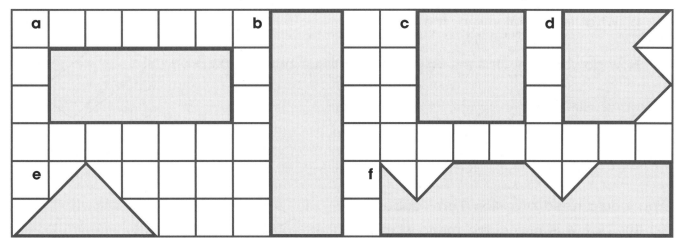

3 Estimate the area of each object. Write the answer beside each shape.

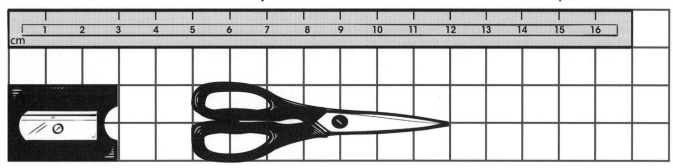

Extension: How many items can you find in your classroom that measure 5 cm²? List them. _____

Unit 4

Length, Area and Volume (TRB pp. 32–35)
Area MA3-10MG selects and uses the appropriate unit to calculate areas, including areas of squares, rectangles and triangles

17

Small and Large Volumes

1 These models are made from centimetre cubes. Find the volume of each model.

a

b

c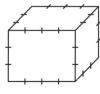

_____   _____   _____

2 Draw in the centimetre cubes and find the volume of each prism.

a   **b**   **c**

_____   _____   _____

3 Look at the dimensions of this box.

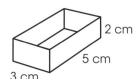

    a How many 1 cm³ cubes would fit on the bottom layer? _____

    b How many layers would there be? _____

    c What is the volume of the box? _____

4 How do you know that the volume of this tissue box is 2000 cm³?

_____

_____

_____

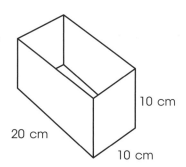

You would need a million 1 cm³ cubes to
fill this box. It is one cubic metre (1 m³).

5 What is the volume of this
store room?

_____

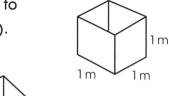

18 **Unit 4** **Length, Area and Volume** (TRB pp. 32–35)
Volume and capacity MA3-11MG selects and uses the appropriate unit to estimate, measure and calculate volumes and
capacities, and converts between units of capacity

STUDENT ASSESSMENT

You will need: a ruler

1 Tick the most appropriate unit of measurement for each item.

	Item	mm	cm	m	km
a	Distance of a marathon				
b	Thickness of a ruler				
c	Length of a basketball court				

2 Measure each line to the nearest mm and cm.

a _____

b _____

c _____

3 Find the area of each shape. Write the answer inside each shape.

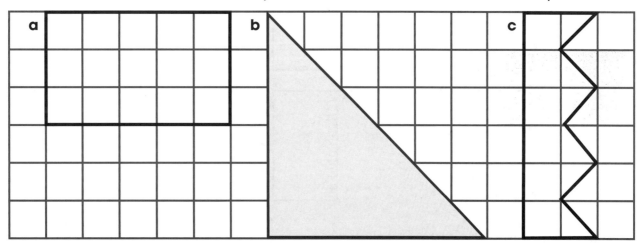

4 Find the volume of each object (**a** and **b** are made from centimetre cubes).

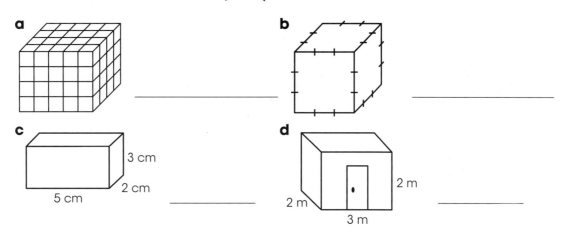

a _____

b _____

c 3 cm / 2 cm / 5 cm _____

d 2 m / 2 m / 3 m _____

Unit
4

Length, Area and Volume (TRB pp. 32–35)
Length MA3-9MG selects and uses the appropriate unit and device to measure lengths and distances, calculates perimeters, and converts between units of length

Area MA3-10MG selects and uses the appropriate unit to calculate areas, including areas of squares, rectangles and triangles

Volume and capacity MA3-11MG selects and uses the appropriate unit to estimate, measure and calculate volumes and capacities, and converts between units of capacity

19

Capacity

You will need: coloured pencils

1 Tick the most appropriate unit of measurement for the capacity of each item.

	Object	mL	L
a	A coffee cup		
b	A bucket		
c	A medicine cup		
d	A swimming pool		
e	A fish tank		
f	A glass		

2 Colour each jug to show the given capacity.

a 500 mL

b 250 mL

c 800 mL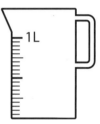

d $\frac{1}{2}$ L

e $\frac{3}{4}$ L

f $\frac{1}{4}$ L

3 Write the amount in each container.

a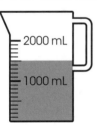

b

c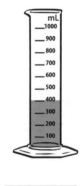

_____   _____   _____

Extension: Order the capacities from **least** to **greatest**.

$2\frac{1}{2}$ L   2 300 mL   200 mL   2 L   2 L 450 mL   22 000 mL

_____

  Mass and Capacity (TRB pp. 36–39)
Volume and capacity MA3-11MG selects and uses the appropriate unit to estimate, measure and calculate volumes and capacities, and converts between units of capacity

Capacity and Volume

You will need: coloured pencils

1 Students collected and measured rainfall each day for a week. The recorded quantities are shown on the rain gauges.

Monday   Tuesday   Wednesday   Thursday   Friday   Saturday   Sunday

a Record how much rain fell each day.

   M _____ T _____ W _____ Th _____ F _____ S _____ Sun _____

b On which day did the **most** rain fall? _____

c What was the **least** amount of rain collected in 1 day? _____

d What was the total rainfall for the week? _____

2 Students collected rainfall during the following week. The recorded quantities are shown below the rain gauges.

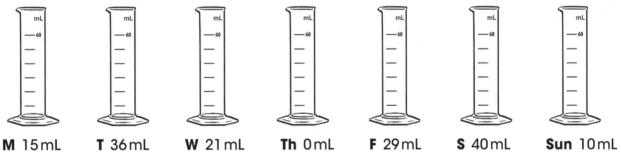

M 15 mL   **T** 36 mL   **W** 21 mL   **Th** 0 mL   **F** 29 mL   **S** 40 mL   **Sun** 10 mL

a Colour each rain gauge to show the rainfall for that day.

b Write 3 questions you could ask about this set of rain gauges.

   i _____

   ii _____

   iii _____

Extension: If 1 cubic centimetre equals 1 millilitre, find the capacity in millilitres of each of the following.

   a 25 cm³ _____   **b** 150 cm³ _____   **c** 256 cm³ _____   **d** 495 cm³ _____

Unit 5
Mass and Capacity (TRB pp. 36–39)
Volume and capacity MA3-11MG selects and uses the appropriate unit to estimate, measure and calculate volumes and capacities, and converts between units of capacity

21

Mass

1 Tick the most appropriate unit of measurement for the mass of each item.

	Item	Gram (g)	Kilogram (kg)	Tonne (t)
a	A snail			
b	A brick			
c	An aeroplane			
d	A train			
e	A box of fruit			
f	A sheet of paper			

2 If 1 kilogram equals 1 thousand grams, how many grams are in each of the following?

a 4 kg _____

b 9 kg _____

c 3.5 kg _____

d 6.2 kg _____

e 5 kg 320 g _____

f $1\frac{1}{4}$ kg _____

3 How many of each item can be packed into a box that can hold a mass of 5 kg?

a

500 g _____

b

100 g _____

c

250 g _____

d

2 kg _____

e

1.5 kg _____

f

200 g _____

4 Use a greater than (>) or less than (<) symbol to complete each statement.

a 417 g ☐ 0.5 kg

b 2500 g ☐ 2 kg

c 2100 g ☐ 2 kg

d 150 g ☐ 0.5 kg

Extension: Do some research to find the **heaviest** animal on Earth.

On another sheet of paper, write a description, including details about its weight.

5 STUDENT ASSESSMENT

1 Tick the most appropriate unit of measurement for the items.

	Item	mL	L	cm³	g	kg
a	Amount of sauce in a bottle					
b	The number of items that would fit in a fruit box					
c	The difference in mass between 2 pieces of fruit					
d	The amount of medicine to give a sick child					
e	The amount of water in a bucket					
f	The mass of a stack of books					

2 Write the amount in each container.

a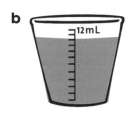
2000 mL
1000 mL

_____

b 12 mL

_____

c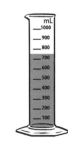
mL
1000
900
800
700
600
500
400
300
200
100

_____

3 Write the mass of each object.

a
500 600 700
400 800
300 900
200 1000
100 g 1100
0

_____

b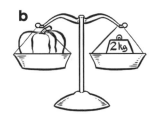
2 kg

_____

c
0 1 2 3 4 5 6 7 8 9

_____

4 Circle the **larger** amount.

a 2 kg or 2300 g   **b** 50 mL or 5 L

c 150 g or 1.5 kg   **d** 5 L or 4 L 200 mL

e 40 kg or 4000 g   **f** 70 mL or 0.7 L

5 Explain what each word means.

a mass _____

b capacity _____

Unit
5
Mass and Capacity (TRB pp. 36–39)
Volume and capacity MA3-11MG selects and uses the appropriate unit to estimate, measure and calculate volumes and capacities, and conver ts between units of capacity

Mass MA3-12MG selects and uses the appropriate unit and device to measure the masses of objects, and conver ts between units of mass

23

Rounding

DATE:

1 Round each number to the nearest ten.

    a 52 _____   **b** 18 _____

    c 79 _____   **d** 102 _____

    e 126 _____   **f** 375 _____

    g 498 _____   **h** 6301 _____

2 Round each number to the nearest hundred.

    a 111 _____   **b** 408 _____

    c 354 _____   **d** 956 _____

    e 1181 _____   **f** 209 _____

    g 2949 _____   **h** 6147 _____

3 Round each number to the nearest thousand.

    a 1040 _____   **b** 2714 _____

    c 3978 _____   **d** 7156 _____

    e 4358 _____   **f** 12724 _____

    g 16700 _____   **h** 21467 _____

4 Circle the numbers that round to 2700, which is the nearest hundred.

    2714   2439   2769   2695   2715   2651

5 Write 5 numbers that would round to 3100, which would be their nearest hundred.

    _____

    _____

Estimation (TRB pp. 40–43)
Whole numbers MA3-4NA orders, reads and represents integers of any size and describes properties of whole numbers

Estimation

1 Round each number to the **nearest ten** and then complete the estimate total.

a 152 + 49 = _____ + _____ E = _____

b 409 + 358 = _____ + _____ E = _____

c 184 + 323 = _____ + _____ E = _____

d 189 − 52 = _____ − _____ E = _____

e 523 − 79 = _____ − _____ E = _____

f 426 − 83 = _____ − _____ E = _____

2 Round each number to the **nearest hundred** and then complete the estimate total.

a 314 + 489 = _____ + _____ E = _____

b 791 + 532 = _____ + _____ E = _____

c 866 + 419 = _____ + _____ E = _____

d 992 − 479 = _____ − _____ E = _____

e 615 − 294 = _____ − _____ E = _____

3 Round the numbers in the box to the **nearest thousand**. Write the rounded pairs that, when added, total the numbers below.

| 4 190 | 2 950 | 4 650 | 7 752 | 9 325 | 1 168 | 6 925 |

a 7 000 _____ **b** 15 000 _____ **c** 10 000 _____

d 13 000 _____ **e** 8 000 _____ **f** 12 000 _____

Extension: Round the numbers and estimate the product.

a 4 789 × 5 =   **b** 2 406 × 8 =

Unit 6 **Estimation** (TRB pp. 40–43)
Addition and subtraction MA3-5NA selects and applies appropriate strategies for addition and subtraction with counting numbers of any size

25

Using Estimation

1   Round to 10 the amounts in each price tag pair, and total. Draw a line matching your estimated total to a money pile.

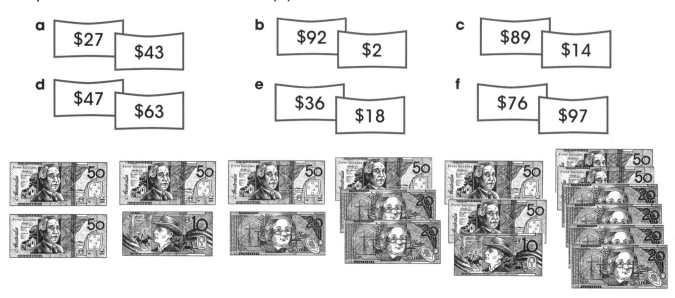

a $27 $43

b $92 $2

c $89 $14

d $47 $63

e $36 $18

f $76 $97

2   Estimate the total cost of the shopping. Select the amount of money that the total matches most closely.

$3.27 $4.19 $2.95 $1.35 $1.07 $4.96

Hint: round to the nearest 10 cents

Extension: Look at a catalogue from a shop, and list 5 items that you would like to buy. Estimate how much money you would need to pay for them.

Unit

6

STUDENT ASSESSMENT

DATE:

1 Round each number to the nearest 10, 100 and 1000.

	Number	Nearest 10	Nearest 100	Nearest 1000
a	1023			
b	2498			
c	1566			
d	4722			
e	7805			
f	9930			

2 Round each number to the nearest 100 and then complete the estimate.

a 715 + 256 = _____ + _____ E = _____

b 481 + 132 = _____ + _____ E = _____

c 1705 + 4562 = _____ + _____ E = _____

d 3992 − 1897 = _____ − _____ E = _____

e 6289 − 1157 = _____ − _____ E = _____

3 Estimate the total of the price tags to the nearest dollar.

\$5.95		\$7.15		\$8.76		\$10.09

_____ _____ _____ _____

4 Complete the equations, and then use estimation to check your answers.

a 465 + 789 =   **b** 4894 + 1138 =

c 9874 − 5698 =   **d** 5460 − 3285 =

Unit

6

Estimation (TRB pp. 40–43)
Whole numbers MA3-4NA orders, reads and represents integers of any size and describes properties of whole numbers
Addition and subtraction MA3-5NA selects and applies appropriate strategies for addition and subtraction with counting
numbers of any size

27

Multiplying Using the Area Model

Shade the arrays into familiar tables to help you complete the multiplication equations.

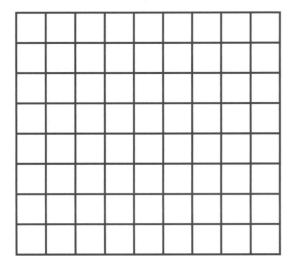

a 9 × 8 =

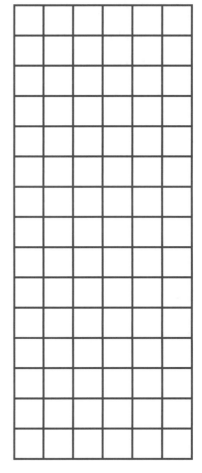

b 6 × 15 =

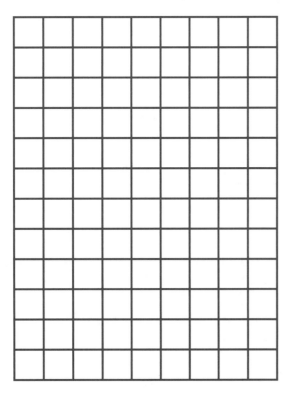

c 9 × 12 =

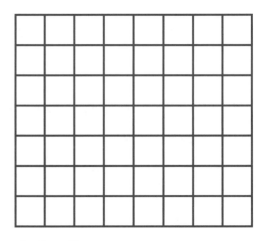

d 8 × 7 =

Multiplication Problems

1   Use the partitioning method to complete the multiplication equations.
    The first one has been started for you.

    a 89 × 8 = (80 × 8) + (9 × 8) =

    b 37 × 3 =

    c 25 × 5 =

    d 77 × 6 =

    e 42 × 7 =

    f 56 × 9 =

2   Complete the equation 74 × 6 = in 3 different ways.

3   Complete the multiplication equations.

    a   7 2   **b**   4 3   **c**   8 1   **d**   9 2
          ×   4   ×   3   ×   6   ×   6
        _____   _____   _____   _____

Extension: If 12 boxes each contain 48 toys,
    how many toys are there altogether?

Unit
7

Multiplication of Large Numbers A (TRB pp. 44–47)
Multiplication and division MA3-6NA selects and applies appropriate strategies for multiplication and
division, and applies the order of operations to calculations involving more than one operation

29

Multiplication of Large Numbers

Complete the multiplication equations. Use the letters that correspond to your answers to solve the puzzle.

89 × 21 =	(Y)	37 × 35 =	(T)
23 × 15 =	(O)	77 × 26 =	(D)
42 × 37 =	(U)	56 × 95 =	(L)
42 × 18 =	(M)	38 × 28 =	(S)
64 × 91 =	(E)	11 × 88 =	(N)
23 × 76 =	(A)	79 × 58 =	(B)

The teacher asked the student: Why are you doing your multiplication on the floor?

The student said:

1869	345	1554	1295	345	5320	2002

756	5824	968	345	1295		1295	345

1554	1064	5824	1295	1748	4582	5320	5824	1064

Multiplication of Large Numbers A (TRB pp. 44–47)
Multiplication and division MA3-6NA selects and applies appropriate strategies for multiplication and division, and applies the order of operations to calculations involving more than one operation

STUDENT ASSESSMENT

DATE:

1 Use the grid to find the answer to 17 × 9.

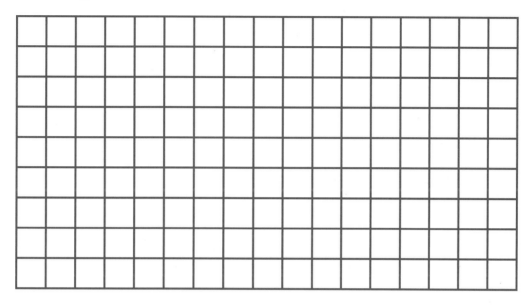

2 Complete the multiplication equation.

45 × 3 = (40 × 3) + (☐ × ☐)

3 Complete the multiplication equations.

a　　3 2
　　× 　4
　　――――

b　　6 7
　　× 　5
　　――――

4 **a** Use your preferred method to complete
the multiplication equation.　　75 × 31

b Explain the method you used in Question 4a.

Extension: Use the lattice
method to complete the
multiplication equation.

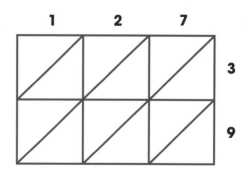

Unit
7
Multiplication of Large Numbers A (TRB pp. 44–47)
Multiplication and division MA3-6NA selects and applies appropriate strategies for multiplication and
division, and applies the order of operations to calculations involving more than one operation

31

Shapes

1 Name each shape.

a  _____

b  _____

c  _____

d  _____

2 Write the name of each type of triangle.

a 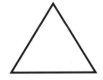 _____

b _____

c _____

d 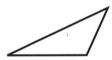 _____

3 Write the number of internal angles (angles inside the shape) for each shape.

a _____

b  _____

c _____

d  _____

4 Write about the similarities and differences between these two shapes.

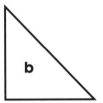

Similarities: _____

Differences: _____

 Unit **8** **Shapes** (TRB pp. 48–51)
Two-dimensional space MA3-15MG manipulates, classifies and draws two-dimensional shapes, including equilateral, isosceles and scalene triangles, and describes their properties

Shapes in Photographs

Look at the photographs.

1 List the different shapes you can see in each.

2 Identify whether the shapes are regular or irregular.

_____

_____

_____

_____

_____

_____

_____

_____

_____

_____

_____

_____

Unit 8 **Shapes** (TRB pp. 48–51)
Two-dimensional space MA3-15MG manipulates, classifies and draws two-dimensional shapes, including equilateral, isosceles and scalene triangles, and describes their properties

33

Drawing 2D Shapes

Complete the table.

a Draw each shape.

b Write a description of each shape.

c Give an example of where each shape might be found in the 'real world'.

Name	Drawing	Description	Example
rectangle			
equilateral triangle			
hexagon			
circle			
kite			
irregular pentagon			
irregular octagon			

Unit 8 **Shapes** (TRB pp. 48–51)
Two-dimensional space MA3-15MG manipulates, classifies and draws two-dimensional shapes, including equilateral, isosceles and scalene triangles, and describes their properties

DATE:

STUDENT ASSESSMENT

You will need: a ruler

1 Name each shape.

a 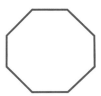 **b**  **c** 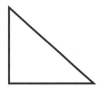 **d**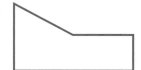

_____ _____ _____ _____

2 Draw the shapes.

    a isosceles triangle   **b** irregular hexagon

3 Describe the shape.

_____

_____

_____

4 List the different shapes
in the photograph.

_____

_____

_____

_____

Shapes (TRB pp. 48–51)
Two-dimensional space MA3-15MG manipulates, classifies and draws two-dimensional shapes, including equilateral, isosceles
and scalene triangles, and describes their properties

35

3D Objects

1 Write the names of the 3D objects in the table.

cube   sphere   triangular prism   rectangular prism   cone   pyramid   cylinder

2 Write a description of each 3D object and an example of where the object might be found in the 'real world'.

Name	Object	Description	Example

3D Objects (TRB pp. 52–55)
Three-dimensional space MA3-14MG identifies three-dimensional objects, including prisms and pyramids, on the basis of their properties, and visualises, sketches and constructs them given drawings of different views

Prisms and Pyramids

1 Name the prisms and pyramids.

a

_____

b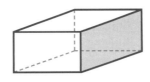

_____

c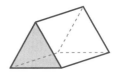

_____

d

_____

e

_____

f

_____

2 Complete the table.

	Object	Number of vertices	Number of edges	Number of faces
a	triangular prism			
b	triangle-based pyramid			
c	rectangular prism			
d	square-based pyramid			
e	hexagonal prism			

3 Draw the set of faces for a hexagonal-based pyramid.

Extension: Name the cross-section of each of the prisms and pyramids.

a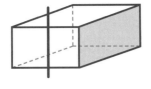

_____

b

_____

c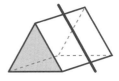

_____

Unit 9 **3D Objects** (TRB pp. 52–55)
Three-dimensional space MA3-14MG identifies three-dimensional objects, including prisms and pyramids, on the basis of their properties, and visualises, sketches and constructs them given drawings of different views

37

Nets

1 Draw the faces that make up the 3D objects.

a   b   c   d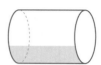

2 Circle the nets that could be used to make each of the 3D objects in Question 1.

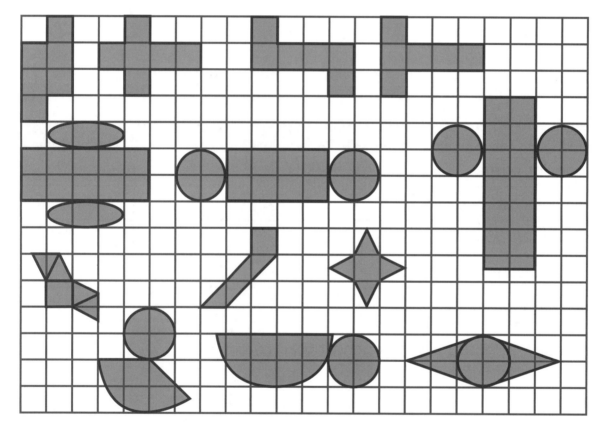

3 Name the 3D object that is made with each net.

a    b    c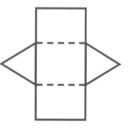

_____   _____   _____

Extension: On another sheet of paper, draw the net for these 3D objects:

a triangular-based pyramid   b rectangular prism

3D Objects (TRB pp. 52–55)
Three-dimensional space MA3-14MG identifies three-dimensional objects, including prisms and pyramids, on the basis of their properties, and visualises, sketches and constructs them given drawings of different views

Unit 9

STUDENT ASSESSMENT

You will need: a ruler

1 Draw each 3D object.

    a cube   **b** cylinder

    c rectangular-based pyramid   **d** triangular prism

2 List the different faces of each 3D object.

    a square-based pyramid _____

    b hexagonal prism _____

3 Draw the net for each 3D object.

    a cube   **b** cylinder

4 Explain the difference between a pyramid and a prism.

_____

_____

Unit 9
3D Objects (TRB pp. 52–55)
Three-dimensional space MA3-14MG *identifies three-dimensional objects, including prisms and pyramids, on the basis of their properties, and visualises, sketches and constructs them given drawings of different views*

39

A Law of Multiplication

You may need: BLM 7 '1 cm Grid Paper'

1 Complete the equations.

a

 =  +

_____ = _____ + _____

b

 = +

_____ = _____ + _____

c

 =  +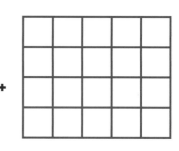

_____ = _____ + _____

2 Complete the equations.

a $4 \times (5 + 6)$ = 4×5 + _____

= 20 + _____

= _____

b $6 \times (3 + 8)$ = 6×3 + _____

= 18 + ____

= _____

c $9 \times (10 + 5)$ = _____ + _____

= _____ + _____

= _____

d $7 \times (5 + 8)$ = _____ + _____

= _____ + _____

= _____

3 Complete the equations.

a 10×16 =

b 9×21 =

Unit 10 **Multiplication of Large Numbers B** (TRB pp. 56–59)
Multiplication and division MA3-6NA selects and applies appropriate strategies for multiplication and division, and applies
the order of operations to calculations involving more than one operation

Multiplication

DATE:

You will need: counters, BLM 7 '1 cm Grid Paper'

1 Use counters set out in arrays to complete the multiplication equations.

a $18 \times 8 =$

b $27 \times 4 =$

c $16 \times 5 =$

d $23 \times 6 =$

2 Use the area model, with 1 cm grid paper, to complete the multiplication equations.

a $22 \times 9 =$

b $47 \times 3 =$

c $18 \times 7 =$

d $28 \times 6 =$

3 Use your preferred method to solve the equations.

a $42 \times 36 =$

b $58 \times 67 =$

4 Describe how you found the answer to Question 3b.

_____

_____

Unit **10** **Multiplication of Large Numbers B** (TRB pp. 56–59)
Multiplication and division MA3-6NA selects and applies appropriate strategies for multiplication and division, and applies the order of operations to calculations involving more than one operation

41

Arthur's Multiplication Problems

You will need: BLM 7 '1 cm Grid Paper', counters

Use grid paper and counters to solve the following questions about Arthur's fruit shop.

a Each cherry tray in Arthur's shop contains 78 cherries. If there are 9 trays altogether, how many cherries does Arthur have?

b Arthur's oranges come in bags of 12.
How many oranges does Arthur have, if he has 48 bags?

c There are 6 tomatoes on each vine, but customers can buy tomatoes individually. If Arthur has 79 vines, how many tomatoes does he have?

d Pineapples come in boxes of 15.
If Arthur has 23 boxes, how many pineapples does he have?

Extension: What is the total amount of fruit in Arthur's fruit shop?

10 STUDENT ASSESSMENT
Unit

Complete the following questions:

1 Use the grid to find 15 × 6.

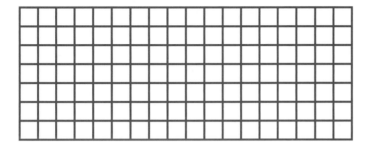

2 Draw an array of counters to verify your answer to Question 1.

3 Complete the equations.

a 9 × (7 + 3) =   **b** 8 × (8 + 4) =

4 Tobi goes to the hardware shop to buy 6 pieces of wood that are each 56 cm long. What is the total length of wood that Tobi has to buy?

5 Write a word problem for 123 × 19 =

_____

_____

Extension: Solve the word problem in Question 5.

Unit
10
Multiplication of Large Numbers B (TRB pp. 56–59)
Multiplication and division MA3-6NA selects and applies appropriate strategies for multiplication and division, and applies the order of operations to calculations involving more than one operation

43

Division

You will need: BLM 5 'Tables Chart 1', BLM 6 'Tables Chart 2'

1 Complete the table. The first row has been started for you.

	Multiplication	Division 1	Division 2
a	6 × 7 =	42 ÷ 6 =	42 ÷ 7 =
b	9 × 8 =		
c	10 × 5 =		
d		24 ÷ 6 =	
e			32 ÷ 4 =
f		27 ÷ 3 =	
g			80 ÷ 10 =

2 Divide each number by 2.

    a 16 _____   **b** 24 _____

    c 18 _____   **d** 10 _____

3 Divide each number by 4.

    a 24 _____   **b** 40 _____

    c 32 _____   **d** 16 _____

4 Complete the division equations.

    a $4\overline{)36}$ **b** $9\overline{)45}$

    c $7\overline{)28}$ **d** $5\overline{)40}$

5 Lily has 48 flowers in a bunch. If she shares the flowers evenly among 4 vases, how many flowers are in each vase?

6 Andrew has 84 counters to share among 4 people. How many counters does each person receive?

Division with Remainders (TRB pp. 60–63)
Multiplication and division MA3-6NA selects and applies appropriate strategies for multiplication and division, and applies the order of operations to calculations involving more than one operation

Daisy's Division

1   Daisy has a selection of plants to plant evenly across 6 garden beds.
    Draw the correct number of plants in each garden bed. How many of
    each plant is left over? _____

    _____

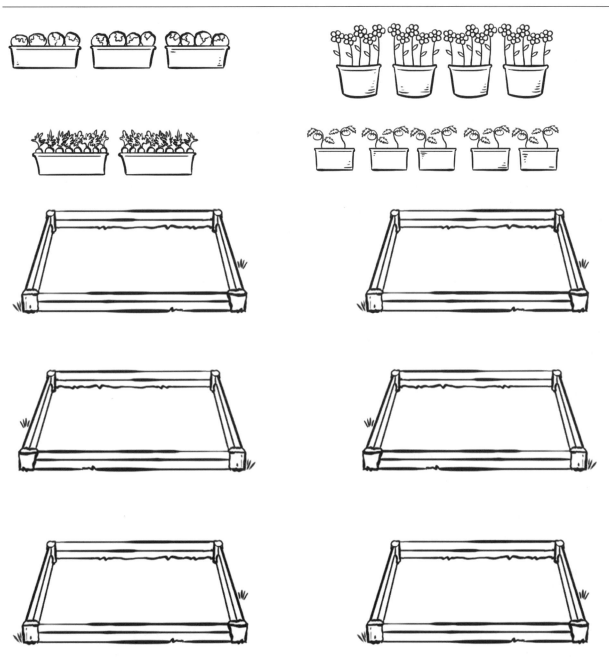

2   Write an equation for each group of plants.

    _____   _____

    _____   _____

Unit 11   **Division with Remainders** (TRB pp. 60–63)
Multiplication and division MA3-6NA selects and applies appropriate strategies for multiplication and division, and applies the order of operations to calculations involving more than one operation

45

Division Word Problems

1   Marcus has been asked to sort the sports equipment
    evenly into 5 sports bins. There are 68 tennis balls,
    10 footballs, 22 basketballs and 15 volleyballs.

    a   How many of each ball is in each bin?

    b   How many balls are left over?

    c   How many balls in total are in each bin?

2   Alex is painting stripes on the basketball court. If 1 tin of paint is used every 5 m,
    how many tins will Alex need to paint 142 m of lines?

3   Penny is making models of ants. Each ant has 6 legs, 2 body parts and 2 feelers.
    Penny has 78 ant legs, 25 body parts and 23 feelers.

    a   How many ant models can Penny make?

    b   How many parts will be left over?

4   Ari's dad is making a wire fence. Each section of the fence is 9 m long.
    The rolls of wire come in 50 m lengths.

    a   How many rolls of wire will he need for the 10 sections of the fence?

    b   How much wire will he have left over?

5   Yellow Valley School needs to transport its 625 students by bus to a
    concert rehearsal. Each bus can hold 30 students.
    How many buses will be needed for the trip?

46   Unit 11   **Division with Remainders** (TRB pp. 60–63)
Multiplication and division MA3-6NA selects and applies appropriate strategies for multiplication and division, and applies the
order of operations to calculations involving more than one operation

DATE:

STUDENT ASSESSMENT

1 Complete the table.

	Multiplication	Division 1	Division 2
a	6 × 9 =		
b	5 × 4 =		
c		36 ÷ 4 =	
d			40 ÷ 8 =

2 Complete the division equations.

a 48 ÷ 6 =   **b** 63 ÷ 9 =

c 49 ÷ 7 =   **d** 56 ÷ 7 =

3 Complete the division equations.

a 9)‾72‾   **b** 4)‾32‾   **c** 3)‾36‾

4 Solve each division problem.

a 50 plastic bricks shared evenly among 4 students

b 73 biscuits shared evenly among 5 plates

c 85 m of ribbon divided evenly into 9 pieces

5 Hugo had 8 pairs of scissors, 16 rulers and 42 pencils. If he shared these items evenly among 4 table groups, how many of each item did each group have?

Extension: Write a word problem for 60 ÷ 7 = 8 r 4

Unit
11
Division with Remainders (TRB pp. 60–63)
Multiplication and division MA3-6NA selects and applies appropriate strategies for multiplication and division, and applies the
order of operations to calculations involving more than one operation

47

One Small Part

Look at the map.

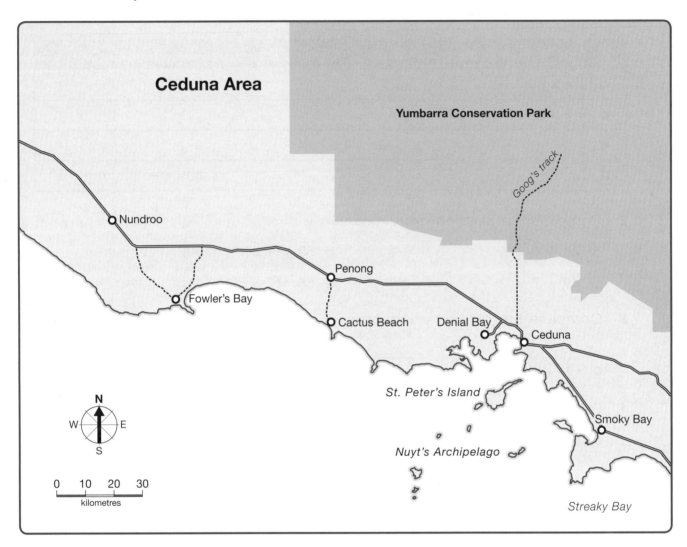

1 List the main landmarks.

_____

2 Describe how you would get from the Yumbarra Conservation Park to Nundroo.

_____

3 Describe where St Peter's Island is located.

_____

4 Investigate the part of Australia that the map illustrates. What other landmarks can be found in this area?

_____

Looking at Western Australia

Look at the map.

1 List the features that can be found at each grid reference.

a E2 _____

b D6 _____

c C4 _____

d E5 _____

2 Find the grid reference for each of the following.

a Perth _____

b Fitzgerald River National Park

_____

c Windjana Gorge National Park

_____

d Broome _____

3 Describe how you would drive from Perth to each destination. Include the directions you would travel and map references.

a Shark Bay _____

_____

b Port Hedland _____

_____

c Kalgoorlie-Boulder _____

_____

Extension: Plan a holiday, starting in Perth. You must visit a beach, a desert and a national park on your trip. On another sheet of paper, name the places you will visit and write directions explaining how to get there. Draw your route on the map above.

Unit 12 **Grid References** (TRB pp. 64–67)
Position MA3-17MG locates and describes position on maps using a grid-reference system

49

Using the Classroom Grid Reference System

You will need: access to the classroom grid reference system you created on large chart paper

Use the classroom grid reference system you created to complete the following.

1   Describe the route from the classroom door to your desk/set. Include the directions and the coordinates. _____

   _____

2   Describe the route from your teacher's desk to your desk/set. Include the directions and the coordinates. _____

   _____

3   Use directions and coordinates to describe a route from one location to another in your classroom. _____

   _____

4   Use directions and coordinates to describe another route from one location to another in your classroom. _____

   _____

5   Reflect on the task of creating your classroom grid reference system.

   a What was difficult about the task? _____

   _____

   b How accurate was the system when you had to complete the descriptions above? _____

   _____

   _____

   c What did you learn from completing the task? _____

   _____

   _____

STUDENT ASSESSMENT

Look at the map.

1 Name and describe the location of 3 landmarks.

a _____

_____

_____

_____

_____

b _____

_____

_____

_____

_____

c _____

_____

_____

_____

_____

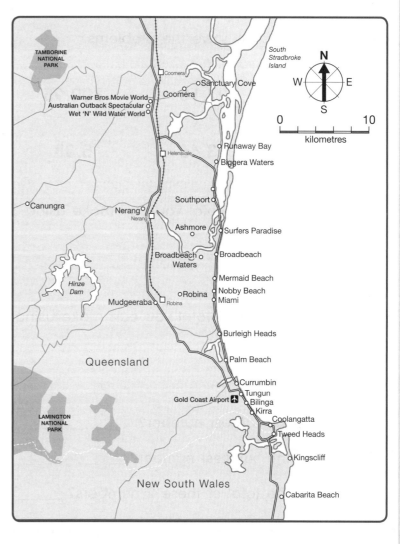

2 Circle 2 landmarks in Question 1. Describe the route between them.

_____

_____

3 Create a grid reference system by drawing over the map. Use 2 cm intervals on your grid.

4 Use your grid reference system to find the grid references of your 3 landmarks in Question 1.

a _____ b _____ c _____

Unit
12
Grid References (TRB pp. 64–67)
Position MA3-17MG locates and describes position on maps using a grid-reference system

51

Adding Decimals

Use a calculator to check your answers!

You will need: a calculator

Use these numbers to solve the problems.

24.01	27.13	300.95	421.5	129.88
74.95	33.46	5.08	105.52	179.3

1 What is the **largest** total you can make using 3 of the numbers?

2 What is the **smallest** total you can make using 3 of the numbers?

3 a What is the largest number? _____

   b What is the smallest number? _____

   c What is the total of these 2 numbers? _____

4 What is the total of all of the numbers less than 100?

5 What is the total of all of the numbers greater than 100?

6 What is the total of all of the numbers?

 Addition and Subtraction of Decimals (TRB pp. 68–71)
Fractions, decimals and percentages MA3-7NA compares, orders and calculates with fractions, decimals and percentages

Find the Missing Shape

1 Complete the subtraction equations.

2 Draw a line matching the shapes that have the **same** answers.

3 Which shape does not have a match? _____

a
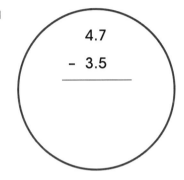
4.7
− 3.5
_____

b
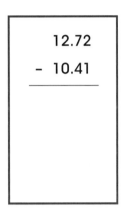
12.72
− 10.41
_____

c
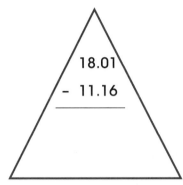
18.01
− 11.16
_____

d
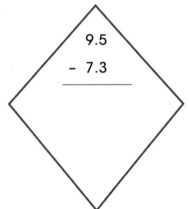
9.5
− 7.3
_____

e
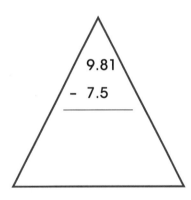
9.81
− 7.5
_____

f
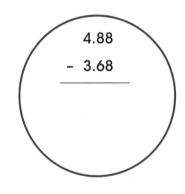
4.88
− 3.68
_____

g
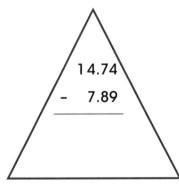
14.74
− 7.89
_____

h
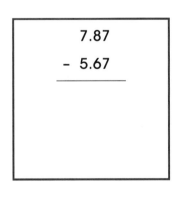
7.87
− 5.67
_____

i
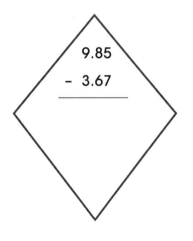
9.85
− 3.67
_____

Unit **13** **Addition and Subtraction of Decimals** (TRB pp. 68–71)
Fractions, decimals and percentages MA3-7NA compares, orders and calculates with fractions, decimals and percentages

53

Addition and Subtraction of Decimals in Context

1 Susanna has 3 lengths of ribbon measuring 1.2 m, 3.5 m and 7.8 m.
What is the total length of ribbon?

2 The weight of 4 button containers is 22.8 g, 19.5 g, 21.7 g and 35.5 g.
What is the total weight of the containers?

3 Peta's mum gave her $8.55 for her lunch, but she lost $5.10 on the way to school.
How much money did she have left?

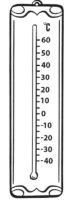

4 Ray recorded the temperature at the beginning of the day as 15.7°C,
and at lunchtime as 25.2°C. What was the increase in temperature?

5 The height of a bookcase is 2.75 m, and the height of a ceiling is 3.28 m.
How much space is there between the top of the bookcase and the ceiling?

6 Laurie collected 2.78 kg of strawberries from his garden on Monday, 3.75 kg on
Tuesday and 1.89 kg on Wednesday. On Thursday, he gave away 2.9 kg to his
friends. What was the mass of strawberries that he had left?

STUDENT ASSESSMENT

Use these numbers to solve the problems.

| 15.25 | 19.07 | 14.73 | 10.08 | 8.45 | 1.28 | 17.79 |

1 Add all of the numbers that are **greater than** 10.

2 Identify the **largest** and the **smallest** number.

3 Find the difference between the **largest** and the **smallest** number.

4 What needs to be added to 8.45 to make 10?

5 What change is given from $20.00 if $17.79 is paid?

6 Write a word problem for 14.73 – 8.45.

_____

_____

7 What is the total of all of the numbers?

8 Explain the strategies you used to find the total in Question 7.

_____

Extension: Two numbers in the list added together give a third number in the list. What are the 3 numbers?

Unit 13

Addition and Subtraction of Decimals (TRB pp. 68–71)
Fractions, decimals and percentages MA3-7NA compares, orders and calculates with fractions, decimals and percentages

55

Perimeter of Rectangles

Don't forget the units!

You will need: a ruler

1   Find the length of each of the sides of the rectangles and label the diagrams.

2   By adding the side lengths, find the total perimeter for each of the rectangles and write this in the centre of the rectangles.

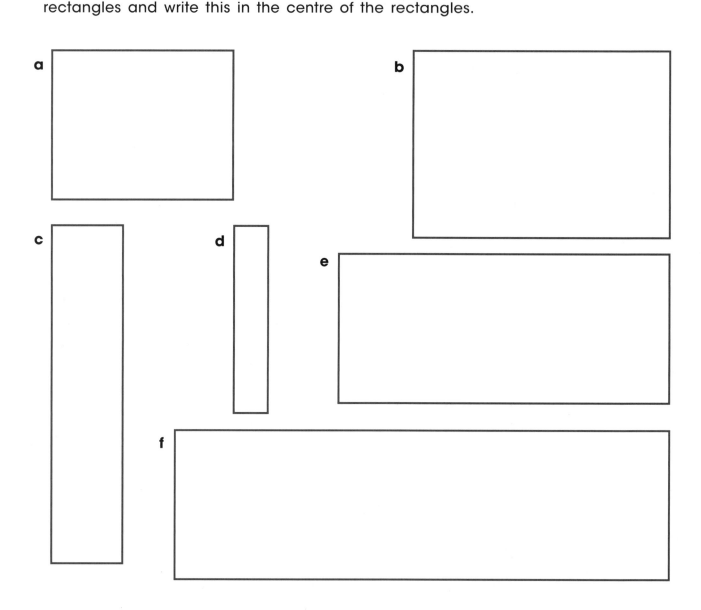

a

b

c

d

e

f

Extension: On another sheet of paper, draw 2 different rectangles that each have a total perimeter of 10 cm.

Rectangles on Grids

Don't forget the units!

You will need: a ruler

1   Find the length of each of the sides of the rectangles and label the diagrams.

2   Find the total perimeter for each of the rectangles and write this in the centre of the rectangles. Use one of these strategies:

   • length + width × 2 = perimeter

   • 2 × length + 2 × width = perimeter

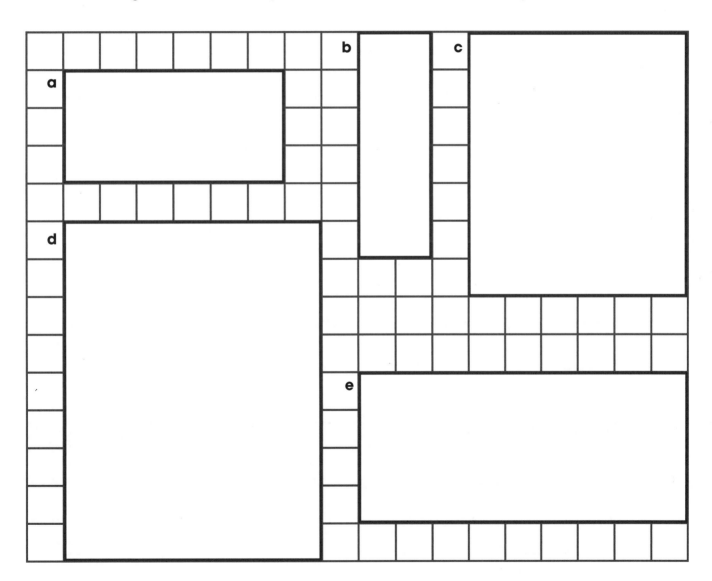

3   Which method did you use to find the perimeter of rectangle **c**?

_____

Unit 14 **Perimeter** (TRB pp. 72–75)
Length MA3-9MG selects and uses the appropriate unit and device to measure lengths and distances, calculates perimeters, and converts between units of length

57

Perimeter

Don't forget the units!

You will need: a ruler

1 Find the perimeter of each shape and label the diagram.

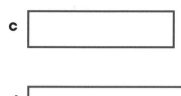

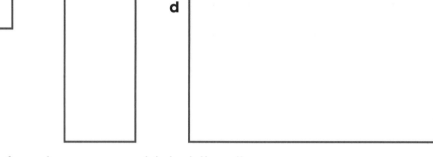

2 Find the perimeter of each square and label the diagram.

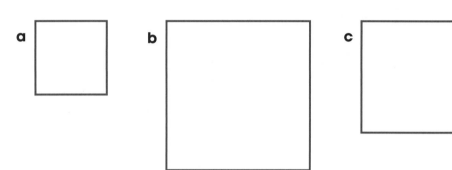

3 Which method did you use to find the perimeter of the square in Question 2c?

_____

4 Complete the table.

Shape	Length	Width	Perimeter
a	7 cm	4 cm	
b	12 cm	10 cm	
c	5 m	1 m	
d	10 m	20 m	

Extension: On another sheet of paper, draw an irregular shape with a perimeter of 20 cm.

STUDENT ASSESSMENT

DATE:

You will need: a ruler

1 Find the missing side lengths of each rectangle and square and add to the labels.

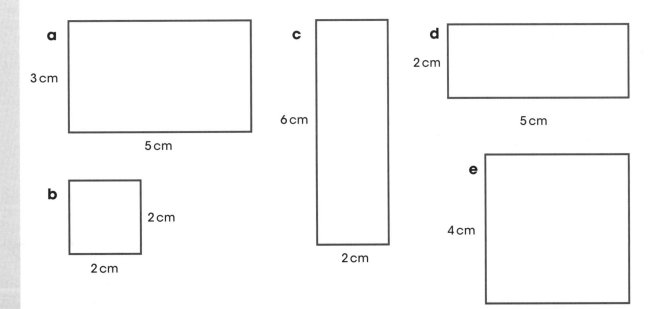

2 Find the perimeter of each rectangle and square. Write the perimeter in each shape.

a — 3 cm, 5 cm

b — 2 cm, 2 cm

c — 6 cm, 2 cm

d — 2 cm, 5 cm

e — 4 cm, 4 cm

3 Explain how you found the perimeter of shape **c** in Question 2.

_____

4 On another sheet of paper, draw a shape that has a total perimeter of 10 cm, and label each of the side lengths.

Unit 14 **Perimeter** (TRB pp. 72–75)
Length MA3-9MG selects and uses the appropriate unit and device to measure lengths and distances, calculates perimeters, and converts between units of length

59

Addition and Subtraction Mental Strategies

1 Write the double of each number.

a 40 _____

b 300 _____

c 900 _____

d 1 200 _____

e 4 200 _____

f 5 000 _____

2 Use doubles to complete the equations. The first 2 have been started for you.

a 520 + 500 = _____ + 20 + 500

b 190 + 160 = _____ + 30 + 160

c 900 + 800 = _____ + _____ + _____

d 385 + 350 = _____ + _____ + _____

e 1 120 + 1 190 = _____ + _____ + _____

f 5 000 + 4 800 = _____ + _____ + _____

3 Write each number in expanded form. Then group like places together and find the answers.

a 750 + 140 =

b 4 341 + 2 605 =

c 769 – 155 =

d 5 609 – 2 401=

4 Use open number lines to complete these equations.

a 192 – 76 =

b 67 + 582 =

c 1 025 – 924 =

d 1 615 + 3 209 =

Remember to start with the largest number first!

5 Explain a strategy you could use to find the answer to 489 + 242.

_____

Multiplication and Division Mental Strategies

1 Complete the equations.

a 4 × 10 =   **b** 10 × 12 =

c 55 × 10 =   **d** 98 × 10 =

e 451 × 10 =   **f** 5000 × 10 =

2 Complete the equations.

a 13 × 5 =   **b** 42 × 5 =

c 106 × 5 =   **d** 502 × 5 =

3 Complete the equations.

a 8 × 20 =   **b** 20 × 35 =

c 56 × 20 =   **d** 711 × 20 =

e 7 × 50 =   **f** 76 × 50 =

g 82 × 50 =   **h** 158 × 50 =

4 Complete the equations.

a 80 ÷ 10 =   **b** 65 ÷ 5 =

c 75 ÷ 5 =   **d** 120 ÷ 10 =

e 150 ÷ 10 =   **f** 1120 ÷ 10 =

g 4160 ÷ 10 =   **h** 4680 ÷ 10 =

5 Complete the equations.

a 98 ÷ 2 =   **b** 616 ÷ 4 =

c 128 ÷ 8 =   **d** 104 ÷ 2 =

e 472 ÷ 4 =   **f** 1000 ÷ 8 =

6 Describe the strategy you used to solve Question 3f.

Unit 15 **Mental Strategies** (TRB pp. 76–79)
Multiplication and division MA3-6NA selects and applies appropriate strategies for multiplication and division, and applies the order of operations to calculations involving more than one operation

61

Mental Strategies

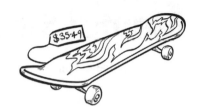

1 Complete the addition equations.

   a 4900 + 5100 + 6200 =

   b 920 + 890 + 240 =

   c 1105 + 1210 + 1320 =

2 Complete the subtraction equations.

   a 825 − 91 =

   b 753 − 328 =

   c 756 − 292 =

3 Use open number lines to complete these equations.

   a 935 + 148 =   **b** 2448 + 279 =

   c 614 − 482 =   **d** 914 − 430 =

4 Find how much change is owed for each item. Do your working out on another sheet of paper.

   a A drink costs \$2.75, and is paid for with \$10.00. _____

   b A basketball costs \$18.55, and is paid for with \$20.00. _____

   c A skateboard costs \$35.49, and is paid for with \$50.00. _____

   d A T-shirt costs \$44.95, and is paid for with \$50.00. _____

5 Find the total of each of the following. Do your working out on another sheet of paper.

   a 5 lengths of rope that are each 1.20 m long _____

   b 10 lengths of wood that are each 1.80 m long _____

   c 100 boxes that each weigh 16 g _____

   d 20 containers that each weigh 56 kg _____

   e 896 kg shared evenly among 8 containers _____

   f 122 m of rope divided evenly into 2 lengths _____

Mental Strategies (TRB pp. 76–79)
Addition and subtraction MA3-5NA selects and applies appropriate strategies for addition and subtraction with counting numbers of any size
Multiplication and division MA3-6NA selects and applies appropriate strategies for multiplication and division, and applies the order of operations to calculations involving more than one operation

STUDENT ASSESSMENT

1 Solve the problems mentally. Write your answer and explain the strategies that you used.

a $342 \times 9 =$

_____

_____

_____

b $196 \div 6 =$

_____

_____

_____

c $7998 + 2001 =$

_____

_____

_____

d $6000 - 3985 =$

_____

_____

_____

2 Here is a division strategy: *to divide any number by 500, multiply by 2 and then divide by 1000.*

a Write an example to show that the strategy works.

b Explain why the strategy works.

_____

3 The local park is holding an outdoor cinema experience. Tickets cost $22 each.

a If 298 people attend, what is the total of the ticket sales?

b Explain how you solved the problem, and the strategies that you used.

_____

Unit
15

Mental Strategies (TRB pp. 76–79)
Addition and subtraction MA3-5NA selects and applies appropriate strategies for addition and subtraction with counting numbers of any size

Multiplication and division MA3-6NA selects and applies appropriate strategies for multiplication and division, and applies the order of operations to calculations involving more than one operation

63

Area of Squares

You will need: a ruler

1 Find the perimeter of each square. Write the perimeter in the shape.

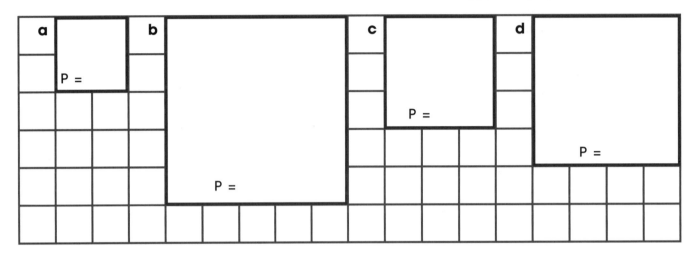

2 Find the area of each square in Question 1.

a _____

b _____

c _____

d _____

3 Measure the side lengths of each square, label the diagram and then find the area of each square.

a

A = _____ cm²

d

A = _____ cm²

b

A = _____ cm²

c

A = _____ cm²

4 Find the area of:

a a square sandpit with 5 m sides. _____

b a square counter with 2 cm sides. _____

c a square sandwich with 15 cm sides. _____

d a square car park with 20 m sides. _____

64 | Unit 16 | **Area** (TRB pp. 80–83)
Length MA3-9MG selects and uses the appropriate unit and device to measure lengths and distances, calculates perimeters, and converts between units of length
Area MA3-10MG selects and uses the appropriate unit to calculate areas, including areas of squares, rectangles and triangles

Area of Rectangles

You will need: a ruler

1 Find the area of each rectangle.

a

A = _____ cm²

b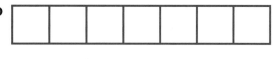

A = _____ cm²

c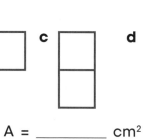

A = _____ cm²

d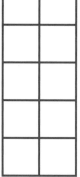

A = _____ cm²

2 Measure the length and width of each rectangle, and write your measurements on the diagram.

a

b

c

d

3 Find the area of each rectangle in Question 2.

a _____

b _____

c _____

d _____

4 Find the area of each rectangle.

a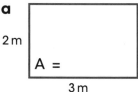
2 m
A =
3 m

b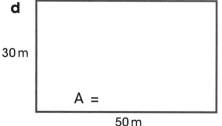
A =
1 m
5 m

c
20 m
A = 10 m

d
30 m
A =
50 m

Extension: Find the area of:

a a rectangle with 6 m and 5 m sides. _____

b a rectangle with 10 cm and 8 cm sides. _____

c a horse paddock 100 m × 160 m. _____

d a house block 80 m × 250 m. _____

Unit 16 **Area** (TRB pp. 80–83)
Length MA3-9MG selects and uses the appropriate unit and device to measure lengths and distances, calculates perimeters, and converts between units of length
Area MA3-10MG selects and uses the appropriate unit to calculate areas, including areas of squares, rectangles and triangles

65

Area of Compound Shapes

You will need: a ruler

1 Find the area of each compound shape. Write the area in the shape.

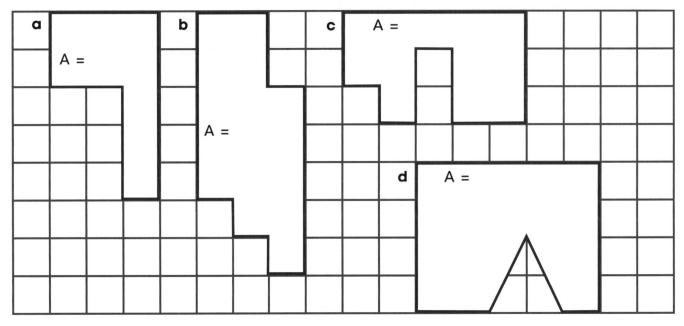

2 Find the area of each compound shape.

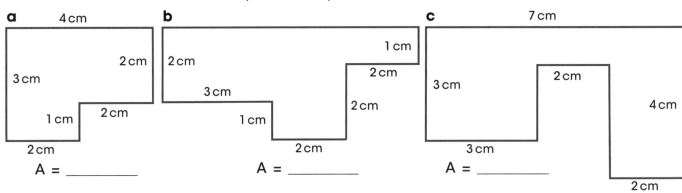

a A = _____

b A = _____

c A = _____

3 Draw a compound shape with an area of 12 cm².

66 Unit 16 **Area** (TRB pp. 80–83)
Length MA3-9MG selects and uses the appropriate unit and device to measure lengths and distances, calculates perimeters, and converts between units of length

Area MA3-10MG selects and uses the appropriate unit to calculate areas, including areas of squares, rectangles and triangles

STUDENT ASSESSMENT

DATE:

1   Find the area of each shape.

a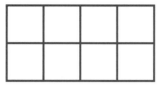

A = _____

b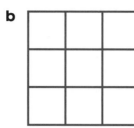

A = _____

c

A = _____

d

A = _____

2   Find the area of each shape.

a

2 m
2 m

A = _____

b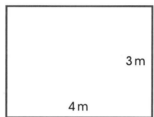

3 m
4 m

A = _____

c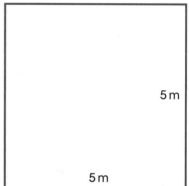

5 m
5 m

A = _____

3   Find the area of each compound shape.

a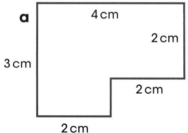

4 cm
2 cm
3 cm
2 cm
2 cm

A = _____

b

1 cm
2 cm   2 cm
2 cm   2 cm
1 cm   1 cm
5 cm

A = _____

c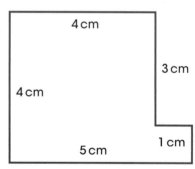

4 cm
3 cm
4 cm
5 cm   1 cm

A = _____

4   Draw a rectangle with an area of 10 cm². Label each of the side lengths.

Extension: Find the area of the triangle.

A = _____

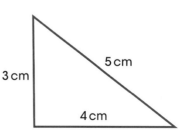

5 cm
3 cm
4 cm

Area (TRB pp. 80–83)
Length MA3-9MG selects and uses the appropriate unit and device to measure lengths and distances, calculates perimeters, and converts between units of length

Area MA3-10MG selects and uses the appropriate unit to calculate areas, including areas of squares, rectangles and triangles

Unit Fractions

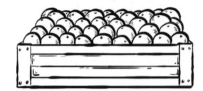

You will need: coloured pencils

1 What fraction of each shape is shaded?

a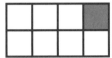

b

c

d

e

2 What fraction of each shape in Question 1 is **not** shaded?

a _____

b _____

c _____

d _____

e _____

3 Shade each shape to match the given fraction.

a $\frac{1}{6}$ b $\frac{1}{4}$ c $\frac{1}{2}$ d $\frac{1}{10}$ 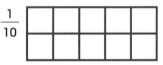 e $\frac{1}{5}$

4 Shade each group to match the given fraction.

a $\frac{1}{3}$ b $\frac{1}{8}$  c $\frac{1}{4}$

d $\frac{1}{5}$ e $\frac{1}{7}$

Extension: Draw a number line from 0 to 2 and divide it into eighths.

Comparing Fractions

DATE:

1 Write the missing fractions on each number line.

a

b

2 Order each set of numbers on a number line.

a $\dfrac{1}{2}$ $\dfrac{1}{4}$ $1\dfrac{1}{4}$ 1

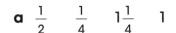

b $\dfrac{1}{5}$ $\dfrac{4}{5}$ $\dfrac{1}{10}$ $\dfrac{1}{2}$ 1

3 Order each set of fractions from **smallest** to **largest**.

a $\dfrac{1}{3}$ $\dfrac{1}{6}$ 1 $\dfrac{2}{3}$ $\dfrac{1}{2}$ _____

b $\dfrac{3}{4}$ $\dfrac{1}{4}$ $\dfrac{1}{3}$ 1 $\dfrac{2}{3}$ _____

c $\dfrac{1}{10}$ $\dfrac{1}{5}$ $\dfrac{1}{2}$ 1 $\dfrac{7}{10}$ _____

d $\dfrac{1}{8}$ $\dfrac{1}{4}$ $\dfrac{1}{2}$ $\dfrac{3}{8}$ $\dfrac{3}{4}$ _____

4 Use < or > to complete the number sentences.

a $\dfrac{1}{2}$ ☐ $\dfrac{1}{3}$ **b** $\dfrac{1}{5}$ ☐ $\dfrac{1}{10}$ **c** $\dfrac{1}{3}$ ☐ $\dfrac{1}{6}$

d $\dfrac{1}{8}$ ☐ $\dfrac{1}{6}$ **e** $\dfrac{1}{10}$ ☐ $\dfrac{1}{2}$ **f** $\dfrac{1}{3}$ ☐ $\dfrac{1}{5}$

Extension: Shade the larger fraction: $\dfrac{1}{4}$ or $\dfrac{5}{12}$.

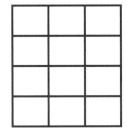

Unit 17 Fractions (TRB pp. 84–87)
Fractions, decimals and percentages MA3-7NA compares, orders and calculates with fractions, decimals and percentages

69

Equivalent Fractions

You will need: coloured pencils

1 Shade the cells alongside each fraction to show an equivalent fraction.
Write the two fractions as an expression. The first one has been done for you.

Fraction	**Equivalent fraction shaded**	**Equation**
$\dfrac{1}{3}$		$\dfrac{1}{3} = \dfrac{2}{6}$
$\dfrac{1}{2}$		
$\dfrac{1}{4}$		
$\dfrac{4}{5}$		
$\dfrac{3}{4}$		
$\dfrac{2}{7}$		
$\dfrac{2}{3}$		
$\dfrac{3}{4}$		
$\dfrac{1}{2}$		
$\dfrac{3}{5}$		
$\dfrac{6}{7}$		

2 Complete the equivalent fractions using the process of simplification.

a $\dfrac{100}{200} = \dfrac{50}{100} = \dfrac{\square}{20} = \dfrac{5}{\square} = \dfrac{\square}{\square}$

b $\dfrac{16}{64} = \dfrac{\square}{32} = \dfrac{4}{\square} = \dfrac{2}{\square} = \dfrac{\square}{\square}$

3 Complete the equivalent fractions.

a $\dfrac{1}{3} = \dfrac{\square}{6} = \dfrac{\square}{9} = \dfrac{10}{\square} = \dfrac{\square}{\square}$

b $\dfrac{2}{5} = \dfrac{\square}{10} = \dfrac{\square}{50} = \dfrac{200}{\square} = \dfrac{\square}{\square}$

STUDENT ASSESSMENT

1 What fraction of each shape is shaded?

a   **b** 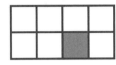   **c**

2 Shade each shape or group the indicated amount.

a $\frac{1}{10}$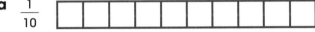

b $\frac{1}{6}$ **c** $\frac{1}{8}$ **d** $\frac{1}{5}$

e $\frac{1}{7}$ **f** $\frac{1}{9}$

3 Order the fractions from **smallest** to **largest**.

a $\frac{1}{4}$ $\frac{1}{3}$ $\frac{1}{2}$ $\frac{2}{3}$ 1 $\frac{3}{4}$ _____

b $\frac{4}{5}$ $\frac{2}{10}$ $\frac{1}{2}$ 1 $\frac{7}{10}$ $\frac{3}{5}$ _____

4 Draw a number line for Question 3a.

5 Write 3 equivalent fractions for each given fraction.

a $\frac{2}{3}$ _____   **b** $\frac{3}{4}$ _____

c $\frac{1}{5}$ _____   **d** $\frac{1}{2}$ _____

Unit **17** **Fractions** (TRB pp. 84–87)
Fractions, decimals and percentages MA3-7NA compares, orders and calculates with fractions, decimals and percentages

71

Decimals to 2 Decimal Places

DATE:

1 Shade each hundreds grid to match the decimal.

a 0.47   **b** 0.63   **c** 0.19   **d** 0.08

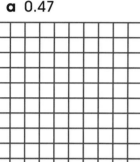 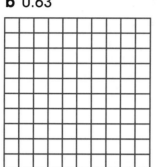 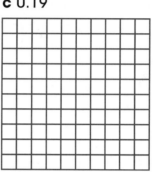 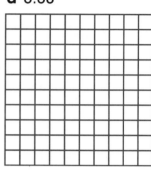

2 Write the numeral shown on each abacus.

a

O · Tth Hth

_____

b

O · Tth Hth

_____

c

O · Tth Hth

_____

3 Write the value of each underlined digit.

a 4.<u>6</u>3 _____   **b** 7.8<u>2</u> _____

c <u>9</u>.46 _____   **d** 10<u>3</u>.78 _____

e <u>5</u>16.23 _____   **f** 732.6<u>6</u> _____

4 Write the decimal number represented by each of the following.

a 2 hundreds, 3 tens, 5 units, 4 tenths and 7 hundredths _____

b 6 hundreds, 5 tens, 8 units, 1 tenth and 9 hundredths _____

c 9 hundreds, 4 tens, 7 units and 5 tenths _____

d 1 hundred, 3 units, 8 tenths and 6 hundredths _____

5 Write how many tenths are in each number.

a 26.89 _____

b 19.46 _____

c 154.11 _____

d 102.39 _____

Extension: Look at the picture.

a Who is the tallest? _____

b Who is the shortest? _____

Chase   Ava   Dom   Kayla

1.28 m   1.23 m   1.11 m   1.38 m

Decimals to 3 Decimal Places

DATE:

1 Write each number in words.

a 7.596 _____

b 11.205 _____

2 Draw the given number on the abacus.

a 4.695   b 7.046   c 17.603

O·Tth Hth Thth   O·Tth Hth Thth   T O·Tth Hth Thth

3 Write the value of each underlined digit.

a 12.459 _____   b 19.867 _____

c 4.221 _____   d 43.986 _____

4 Write the decimal represented by each of the following.

a 5 hundreds, 3 tens, 5 units, 4 tenths, 7 hundredths
and 9 thousandths _____

b 3 hundreds, 8 tens, 8 units, 2 tenths, 9 hundredths
and 1 thousandth _____

c 2 hundreds, 8 tens, 8 units, 5 tenths
and 2 thousandths _____

d 7 hundreds, 3 units, 8 tenths, 4 hundredths
and 5 thousandths _____

5 Circle the **larger** decimal in each pair.

a 24.315   29.428   b 119.46   119.476

c 154.11   155.108   d 102.39   102.039

Extension: Write each of the following as a decimal.

a 22 hundredths _____   b 45 thousandths _____

c 95 hundredths _____   d 46 tenths _____

Unit
18
Decimals to 3 Decimal Places (TRB pp. 88–91)
Fractions, decimals and percentages MA3-7NA compares, orders and calculates with fractions, decimals and percentages

73

Decimals Number Lines

Complete the number lines by filling in the decimals.

a

1.60   1.65

b

10.02   10.03

c

0.14   0.16

d

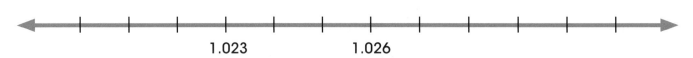

1.023   1.026

e

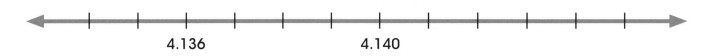

4.136   4.140

f

10.992   10.995

g

10.010

h

22.785

 Unit **18** **Decimals to 3 Decimal Places** (TRB pp. 88–91)
Fractions, decimals and percentages MA3-7NA compares, orders and calculates with fractions, decimals and percentages

DATE:

STUDENT ASSESSMENT

1 Write the numeral shown on each abacus.

a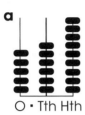

O • Tth Hth

_____

b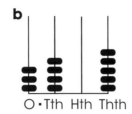

O • Tth Hth Thth

_____

c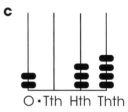

O • Tth Hth Thth

_____

2 Write each number in words.

a 4.76 _____

b 11.03 _____

c 29.063 _____

3 Write the value of each underlined digit.

a 14.<u>6</u>9 _____   **b** 21.4<u>7</u> _____

c 3<u>8</u>.11 _____   **d** 47.08<u>3</u> _____

e 19.<u>1</u>04 _____   **f** 4.8<u>6</u>9 _____

4 Circle the **larger** decimal in each pair.

a 47.11   49.05   **b** 21.39   21.36

c 59.103   59.13   **d** 87.114   87.184

e 112.487   112.874   **f** 324.946   325

5 Circle the measurement that is:

a longer.   4.37 m or 4.385 m

b heavier.   21.46 kg or 22.103 kg

c taller.   1.635 m or 1.621 m

d hotter.   18.72 °C or 18.27 °C

e lighter.   479.3 kg or 479.031 kg

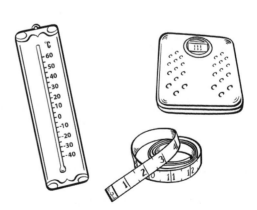

Unit
18
Decimals to 3 Decimal Places (TRB pp. 88–91)
Fractions, decimals and percentages MA3-7NA compares, orders and calculates with fractions, decimals and percentages

75

Chance and Cards

You will need: a standard pack of 52 playing cards

1   Complete the table to show the chance that you would be dealt the cards.
    The first one has been done for you.

Card	Ace	Red card	Spade	5
Chance	1 in 13			
Fraction	$\frac{1}{13}$			

2   What sort of card would you have a 1 in 2 chance
    of being dealt? _____

3   Name a set of cards that you have a 1 in 4 chance of being dealt. _____

4   Name a set of cards that you have $\frac{1}{13}$ of a chance
    of being dealt. _____

5   Two card players are each playing with a different pack of cards. They have been
    dealt the cards shown. Find the chance that the next card dealt to each player is
    a number that falls between the cards shown.

Deal 1	Deal 2
Number of cards left in the pack:	Number of cards left in the pack:
Numbers in between the cards shown:	Numbers in between the cards shown:
Total of these numbers in the pack:	Total of these numbers in the pack:
Chance of it happening:	Chance of it happening:

Extension: What are your 2 favourite cards in a standard pack of 52 playing cards?

_____

a   What chance would you have of these 2 cards being dealt? _____

b   Express this as a fraction. _____

Unit 19   **Chance** (TRB pp. 92–95)
Chance MA3-19SP conducts chance experiments and assigns probabilities as values between 0 and 1 to
describe their outcomes

Spinners

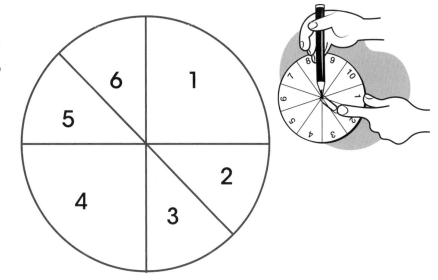

You will need: a pencil with a sharp point, a paperclip

1 Place the paperclip and pencil point in the centre of the spinner.

Spin the paperclip around the pencil.

Collect the tally for 20 spins in the table.

1	2	3	4	5	6

a Which number(s) came up **most** frequently? _____

b Which number(s) came up **least** frequently? _____

c Is it a fair spinner? _____ Why, or why not? _____

_____

d How could the spinner be changed to make it fair? _____

2 Look at the spinners. Record the numbers that are **most** and **least** likely to come up.

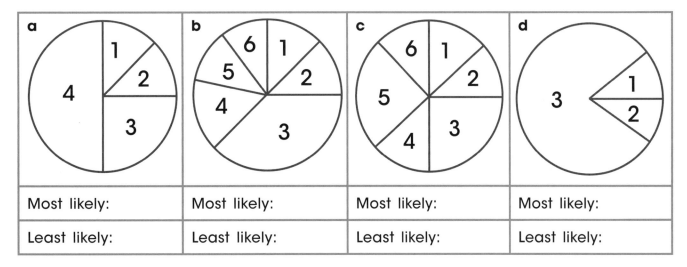

a	b	c	d
Most likely:	Most likely:	Most likely:	Most likely:
Least likely:	Least likely:	Least likely:	Least likely:

Extension: On a sheet of paper, redraw the spinner in Question 1 with the changes you recommended in Question 1d. Is the spinner fair now? _____

Unit 19 **Chance** (TRB pp. 92–95)
Chance MA3-19SP conducts chance experiments and assigns probabilities as values between 0 and 1 to describe their outcomes

77

My Game

1 Select a game that is based on chance. _____

2 List the elements of chance that are involved in the game.

_____

_____

3 List the tool(s) (e.g. a dice) that are used in the game. _____

4 What are the different outcomes involved in the game?

_____

_____

5 Are the chances of the different outcomes the same each time? Explain.

_____

_____

6 Design a game that involves chance.

   a What chance tool(s) will you include? _____

   b What other elements of chance might you include?

_____

_____

   c Describe and draw a picture of your game.

STUDENT ASSESSMENT

1 Draw the possible outcomes of tossing 3 coins.

2 a What does a '1 in 6 chance' mean? _____

_____

b Express a 1 in 6 chance as a fraction. _____

3 What does 'probability' mean? _____

_____

4 a Divide the spinner into 6 sections, numbered 1 to 6, where the sections numbered 2 and 5 are **more likely** to come up.

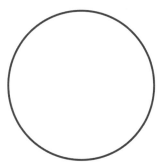

b Is this a fair spinner? _____ Explain. _____

_____

c Redraw the spinner so that it is fair.

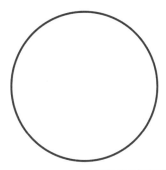

Unit 19   **Chance** (TRB pp. 92–95)
Chance MA3-19SP conducts chance experiments and assigns probabilities as values between 0 and 1 to describe their outcomes

79

My Data Question

Your task is to think of a question, collect the data and organise it.

1 My data question is:

_____

2 The number of people I will survey is: _____

3 Here is the collected data.

4 I have organised it, by: _____

_____

5 Three comments about my data are:

a _____

_____

b _____

_____

c _____

_____

Extension: On another sheet of paper, display your data as a graph.

Written Piece on Data Scenario

Work in a small group, and complete the questions to create a written piece on your collected data.

1 My group decided to collect data on: _____

_____

_____

2 The way we decided to do this was: _____

_____

_____

3 The strength of this was: _____

_____

_____

4 The weakness of this was: _____

_____

_____

5 Our data was valid because: _____

_____

_____

6 What I learnt from collecting the data was: _____

_____

_____

7 If I did this task again, I would like to learn more about: _____

_____

_____

Collecting Data (TRB pp. 96–99)
Data MA3-18SP uses appropriate methods to collect data and constructs, interprets and evaluates data displays, including dot plots, line graphs and two-way tables

Data – School Council Question

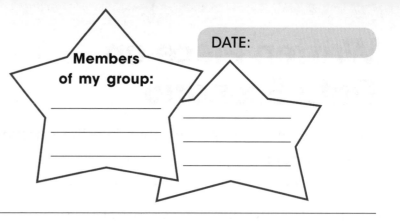

Members of my group:

_____
_____
_____
_____

1 The question we would like to ask the school council is:

_____

_____

2 The way we are going to collect data is:

_____

_____

3 Here is the collected data.

_____

4 What we found from the data is: _____

_____

_____

Unit
20 STUDENT ASSESSMENT

1 List 3 survey questions that could be asked about the topic, 'Pets'.

a _____

b _____

c _____

2 Select 1 of the survey questions from Question 1, and survey at least 10 of your classmates. Collect your data here:

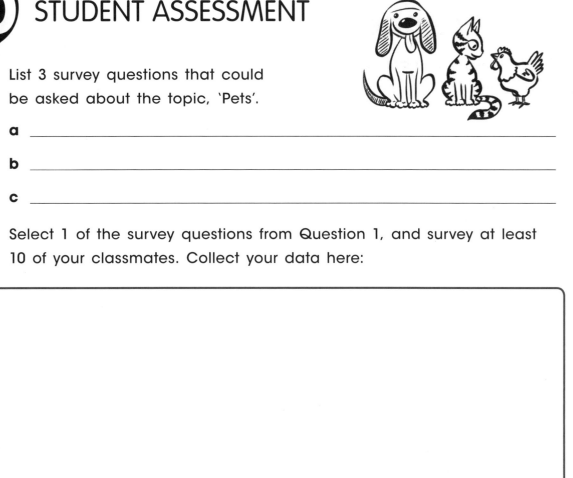

3 Write 3 observations about the data you collected.

a _____

b _____

c _____

4 How could you have made the survey results more valid?

_____

_____

Unit
20 **Collecting Data** (TRB pp. 96–99)
Data MA3-18SP uses appropriate methods to collect data and constructs, interprets and evaluates data displays, including dot plots, line graphs and two-way tables

83

Number Patterns – Whole Numbers

1 Write the next 4 numbers in the number patterns.

a 51, 52, 53, 54, _____, _____, _____, _____

b 22, 24, 26, 28, _____, _____, _____, _____

c 15, 18, 21, 24, _____, _____, _____, _____

d 70, 77, 84, 91, _____, _____, _____, _____

e 103, 106, 109, 112, _____, _____, _____, _____

f 50, 48, 46, 44, _____, _____, _____, _____

g 200, 180, 160, 140, _____, _____, _____, _____

h 64, 56, 48, 40, _____, _____, _____, _____

2 Write the first 5 terms of each number pattern.

a Start at 100 and count forwards by 2s. _____

b Start at 50 and count backwards by 3s. _____

c Start at 109 and count forwards by 1s. _____

d Start at 77 and count backwards by 7s. _____

e Start at 1 000 and count forwards by 10s. _____

f Start at 93 and count backwards by 2s. _____

3 Write a description for 4 of the number patterns in Question 1.

a Question 1_____ : _____

b Question 1_____ : _____

c Question 1_____ : _____

d Question 1_____ : _____

4 Record the 10th term for each number pattern.

a 4, 8, 12, 16, … _____   **b** 1 400, 1 500, 1 600, 1 700, … _____

c 10, 20, 30, 40, … _____   **d** 90, 87, 84, 81, … _____

e 111, 122, 133, 144, … _____   **f** 215, 200, 185, 170, … _____

5 Describe how you found the answer to Question 4f.

_____

_____

Unit 21
Number Patterns (TRB pp. 100–103)
Patterns and algebra MA3-8NA analyses and creates geometric and number patterns, constructs and completes
number sentences, and locates points on the Cartesian plane

Number Patterns – Fractions

DATE:

1 Draw a line matching each pattern to its rule.

a $\frac{1}{2}$ 1 $1\frac{1}{2}$ 2 ... $+ \frac{1}{10}$

b 1 $1\frac{1}{4}$ $1\frac{1}{2}$ $1\frac{3}{4}$ 2 ... $- \frac{1}{4}$

c $\frac{5}{10}$ $\frac{6}{10}$ $\frac{7}{10}$ $\frac{8}{10}$ ... $+ \frac{1}{2}$

d 5 $4\frac{1}{2}$ 4 $3\frac{1}{2}$ ... $- \frac{1}{10}$

e 11 $10\frac{3}{4}$ $10\frac{1}{2}$ $10\frac{1}{4}$ ... $- \frac{1}{2}$

f 1 $\frac{9}{10}$ $\frac{8}{10}$ $\frac{7}{10}$ ... $+ \frac{1}{4}$

2 Write the first 5 terms of each number pattern.

a Start at 5 and count forwards by a $\frac{1}{2}$. _____

b Start at 16 and count backwards by a $\frac{1}{2}$. _____

c Start at $\frac{1}{10}$ and count forwards by $\frac{2}{10}$ s. _____

d Start at 10 and count backwards by $\frac{3}{4}$ s. _____

e Start at $\frac{5}{8}$ and count forwards by an $\frac{1}{8}$. _____

f Start at 1 and count backwards by a $\frac{1}{3}$. _____

3 Write the next 4 numbers in the number patterns.

a $\frac{7}{9}$ 1 $1\frac{2}{9}$ $1\frac{4}{9}$ _____

b $2\frac{1}{10}$ 2 $1\frac{9}{10}$ $1\frac{8}{10}$ _____

c $1\frac{5}{8}$ 2 $2\frac{3}{8}$ $2\frac{6}{8}$ _____

d 5 $4\frac{4}{5}$ $4\frac{3}{5}$ $4\frac{2}{5}$ _____

4 Write number patterns by **adding** a fraction to the starting numbers.
Write 5 numbers in your new patterns.

a $\frac{1}{3}$ _____

b $2\frac{3}{4}$ _____

c $10\frac{1}{5}$ _____

d $\frac{9}{10}$ _____

5 List some examples where fraction number patterns may be used in real life.

_____

Unit **21** **Number Patterns** (TRB pp. 100–103)
Patterns and algebra MA3-8NA analyses and creates geometric and number patterns, constructs and completes number sentences, and locates points on the Cartesian plane

85

Number Patterns – Decimals

You will need: coloured pencils or felt pens

1 **a** Create a colour key by shading the box at the start of each number pattern in a different colour.

b Locate each number pattern in the grid, and shade the whole sequence of numbers in that pattern using the colours from your key.

☐	**1**	5.1, 5.2, 5.3, 5.4, …	☐	**8** 4.5, 4.45, 4.4, 4.35, …
☐	**2**	3.5, 3.7, 3.9, 4.1, …	☐	**9** 3.9, 4, 4.1, 4.2, …
☐	**3**	9.2, 9, 8.8, 8.6, …	☐	**10** 8.6, 8.9, 9.2, 9.5, …
☐	**4**	0.6, 1.2, 1.8, 2.4, …	☐	**11** 3.11, 3.22, 3.33, …
☐	**5**	0.5, 1.0, 1.5, 2.0, …	☐	**12** 5.3, 5.31, 5.32, …
☐	**6**	0.9, 1.8, 2.7, 3.6, …	☐	**13** 1.2, 2.4, 3.6, …
☐	**7**	3.7, 3.75, 3.8, 3.85, …	☐	**14** 3.6, 3.1, 2.6, 2.1, …

Use the unshaded numbers to find the 6-digit code. ☐ ☐ ☐ ☐ ☐ ☐

3.5	4.3	4.35	4.4	4.45	4.5	2.1	3.9	9.2
3.7	3.75	3.8	3.85	3.9	4	2.6	4	9
3.9	3	1	0.9	1.2	3.5	3.1	4.1	8.8
4.1	0.6	1.2	1.8	2.4	3	3.6	4.2	8.6
4.3	3.55	5.35	2.7	3.6	2.5	9.8	4.3	8.4
4.5	3.44	5.34	3.6	4.8	2	9.5	4.4	8.2
4.7	3.33	5.33	4.5	6.0	1.5	9.2	4.5	8
4.9	3.22	5.32	5.4	7.2	1	8.9	4.6	7.8
9	3.11	5.31	6.3	8.4	0.5	8.6	4.7	7.6
5.1	5.2	5.3	5.4	5.5	5.6	5.7	5	8.2

Hint: patterns go forwards, backwards, vertically and horizontally, but not diagonally.

STUDENT ASSESSMENT

1 Write the next 4 numbers in the number patterns.

a 19, 24, 29, 34, _____, _____, _____, _____

b 180, 177, 174, 171, _____, _____, _____, _____

c $6\frac{1}{10}$, $6\frac{3}{10}$, $6\frac{5}{10}$, $6\frac{7}{10}$, _____, _____, _____, _____

d $5\frac{3}{5}$, $5\frac{1}{5}$, $4\frac{4}{5}$, $4\frac{2}{5}$, _____, _____, _____, _____

e 5.16, 5.19, 5.22, 5.25, _____, _____, _____, _____

f 21.06, 21.05, 21.04, 21.03, _____, _____, _____, _____

2 Write the first 5 terms of each number pattern.

a Start at 90 and count forwards by 11s. _____

b Start at 365 and count backwards by 6s. _____

c Start at 21 and count forwards by an $\frac{1}{8}$. _____

d Start at 19 and count backwards by a $\frac{1}{3}$. _____

e Start at 114.5 and count forwards by 0.3. _____

f Start at 50 and count backwards by 1.2. _____

3 For each topic, create a number pattern and write a description of it.

a whole numbers _____

_____

b fractions _____

_____

c decimals _____

_____

Extension: On another sheet of paper, draw the number pattern
21.06, 21.05, 21.04, 21.03, ... on a number line.
Continue the sequence for 5 more numbers.

Unit
21
Number Patterns (TRB pp. 100–103)
Patterns and algebra MA3-8NA analyses and creates geometric and number patterns, constructs and completes
number sentences, and locates points on the Cartesian plane

87

Angles

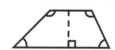

1 Name each type of angle.

a

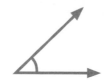

b

c

d

e

f

2 True or false (T or F)?

a 125° is a right angle _____   **b** 270° is a straight angle _____

c 35° is an acute angle _____   **d** 200° is a reflex angle _____

e 89° is an obtuse angle _____   **f** 360° is a revolution _____

3 Draw an angle that is:

a less than 90°.   **b** greater than 180°.

c exactly 90°.   **d** exactly 180°.

e greater than 180°
but less than 270°.   **f** less than 360° and
greater than 180°.

Extension: Name the different types of angles in the shapes.

a

b

c

Measuring Angles

You will need: a protractor, a ruler

1 Record the size of each angle.

a

b

c

d

e

f

2 Estimate the size of each angle.

a

b

c

d

e

f

3 Use a protractor to measure each angle in Question 2.

a _____   **b** _____

c _____   **d** _____

e _____   **f** _____

Extension: Use a protractor to draw each angle.

a 80°   **b** 110°

Unit 22 **Angles** (TRB pp. 104–107)
Angles MA3-16MG measures and constructs angles, and applies angle relationships to find unknown angles

89

Constructing Angles

You will need: a protractor, a ruler

1 a Draw an example of each angle.

   b Measure each angle with a protractor and include the measurement.

	Acute angle	Obtuse angle	Reflex angle
Example of angle			
Angle measurement			

2 Use a protractor to measure each angle.

a

b

c

3 Use a protractor to draw each angle.

   a 70°   **b** 135°   **c** 250°

Extension: Is it possible to draw a shape with one 45° angle and three 100° angles? Explain your answer. _____

_____

STUDENT ASSESSMENT

You will need: a ruler, a protractor

1 Draw an angle that is:

    a less than 90°.

    b greater than 90° but less than 180°.

    c greater than 180° but less than 360°.

    d exactly 90°.

2 Name each of the angles in Question 1 (e.g. straight angle).

    a _____

    b _____

    c _____

    d _____

3 Use a protractor to draw each angle.

    a 260°   **b** 30°

4 Use a protractor to measure each angle.

    a    **b**

Unit 22 **Angles** (TRB pp. 104–107)
Angles MA3-16MG measures and constructs angles, and applies angle relationships to find unknown angles

91

Addition of Fractions

1 Colour each diagram to find the answers to the equations.

a $\frac{3}{10} + \frac{5}{10} =$ **b** $\frac{1}{4} + \frac{1}{4} =$ **c** $\frac{1}{5} + \frac{2}{5} =$

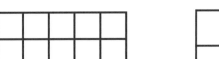

d $\frac{3}{6} + \frac{1}{6} =$ **e** $\frac{3}{8} + \frac{2}{8} =$ **f** $\frac{1}{3} + \frac{1}{3} =$

2 Draw a diagram and complete the equations.

a $\frac{3}{5} + \frac{1}{5} =$ **b** $\frac{1}{7} + \frac{5}{7} =$ **c** $\frac{4}{8} + \frac{3}{8} =$

d $\frac{2}{4} + \frac{1}{4} =$ **e** $\frac{7}{10} + \frac{2}{10} =$ **f** $\frac{3}{6} + \frac{2}{6} =$

3 Complete the equations.

a $\frac{5}{8} + \frac{1}{8} =$ **b** $\frac{1}{4} + \frac{2}{4} =$

c $\frac{1}{5} + \frac{4}{5} =$ **d** $\frac{1}{10} + \frac{7}{10} =$

4 Circle the pairs of fractions that when added together give the answer.

a $\frac{3}{6}$ $\frac{2}{3}$ $\frac{2}{6}$ $= \frac{5}{6}$ **b** $\frac{4}{8}$ $\frac{1}{8}$ $\frac{3}{8}$ $= \frac{5}{8}$

Addition of Fractions and Number Lines

1 Complete the number line.

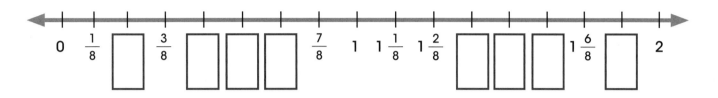

2 Use the number line to answer the equations.

a $1 + \dfrac{1}{4} =$

b $\dfrac{1}{4} + \dfrac{5}{4} =$

c $1 + 1\dfrac{1}{4} =$

d $\dfrac{3}{4} + \dfrac{3}{4} =$

e $\dfrac{2}{4} + \dfrac{5}{4} =$

f $\dfrac{3}{4} + \dfrac{5}{4} =$

3 Use the number line to answer the equations.

a $\dfrac{2}{5} + \dfrac{4}{5} =$

b $1\dfrac{1}{5} + \dfrac{3}{5} =$

c $\dfrac{4}{5} + \dfrac{6}{5} =$

d $1\dfrac{2}{5} + \dfrac{3}{5} =$

4 Complete the equations.

a $\dfrac{3}{8} + \dfrac{4}{8} =$

b $1\dfrac{1}{8} + \dfrac{3}{8} =$

c $\dfrac{4}{10} + \dfrac{5}{10} =$

d $1\dfrac{3}{10} + \dfrac{5}{10} =$

e $\dfrac{3}{5} + \dfrac{1}{5} =$

f $1\dfrac{1}{5} + \dfrac{2}{5} =$

Unit 23 **Addition of Fractions** (TRB pp. 108–111)
Fractions, decimals and percentages MA3-7NA compares, orders and calculates with fractions, decimals and percentages

93

Fraction Addition Problems

Solve the problems.

1 Dad added $\frac{1}{8}$ of a cup of choc bits and $\frac{4}{8}$ of a cup of dried fruit to a cake mixture. How much did he add altogether?

2 Yesterday, I collected $\frac{1}{10}$ of a cup of water from a leaking tap. Today, I collected $\frac{2}{10}$ of a cup from the same leaking tap. How much water did I collect?

3 Stella's new puppy had to be fed $\frac{3}{6}$ of a cup of wet dog food and $\frac{2}{6}$ of a cup of dry dog food every day. How much food did Stella give her puppy each day?

4 In cooking class, Leo received $\frac{1}{10}$ of the cake and Zac received $\frac{2}{10}$. How much cake did the boys receive altogether?

5 Olivia collected $\frac{4}{10}$ of the fallen leaves and her sister collected $\frac{5}{10}$. How much of the fallen leaves did they collect?

Extension: What fraction of the leaves still needed to be collected?

6 In a packet of felt pens, $\frac{3}{10}$ of the pens did not work at all and $\frac{2}{10}$ worked only faintly. What fraction of the felt pens did not work properly?

Extension: What fraction of the felt pens did work properly?

Unit 23
STUDENT ASSESSMENT

You will need: coloured pencils or felt pens

1 Colour each diagram to find the answers to the equations.

a $\frac{6}{10} + \frac{3}{10} =$

b $\frac{5}{8} + \frac{3}{8} =$

c $\frac{1}{4} + \frac{2}{4} =$

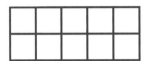

2 Complete the equations.

a $\frac{4}{6} + \frac{1}{6} =$

b $\frac{2}{5} + \frac{1}{5} =$

c $\frac{4}{12} + \frac{6}{12} =$

d $\frac{7}{10} + \frac{2}{10} =$

e $1\frac{1}{3} + \frac{1}{3} =$

f $1\frac{4}{8} + \frac{3}{8} =$

3 Use the number line to solve the equations.

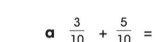

a $\frac{3}{10} + \frac{5}{10} =$

b $\frac{9}{10} + \frac{6}{10} =$

c $1\frac{2}{10} + \frac{5}{10} =$

d $\frac{8}{10} + \frac{9}{10} =$

4 A man had $\frac{2}{6}$ of a metre of one type of rope and $\frac{5}{6}$ of a metre of another type of rope.
What length of rope did he have altogether?

Extension: Write your own word problem for the equation: $\frac{7}{10} + \frac{8}{10} =$

_____

Unit 23 **Addition of Fractions** (TRB pp. 108–111)
Fractions, decimals and percentages MA3-7NA compares, orders and calculates with fractions, decimals and percentages

95

Subtraction of Fractions

1 Complete the equations, using the diagrams to help you.

a $\frac{3}{4} - \frac{1}{4} =$

b $\frac{9}{10} - \frac{6}{10} =$

c $\frac{6}{8} - \frac{3}{8} =$

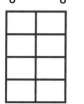

d $\frac{2}{3} - \frac{1}{3} =$

e $\frac{5}{6} - \frac{2}{6} =$

f $\frac{3}{5} - \frac{1}{5} =$

2 Draw a diagram and complete the equations.

a $\frac{7}{10} - \frac{2}{10} =$

b $\frac{5}{12} - \frac{1}{12} =$

c $\frac{4}{5} - \frac{3}{5} =$

d $\frac{7}{8} - \frac{6}{8} =$

3 Complete the equation and show the answer on the diagram.

$3 - \frac{3}{4} =$

4 Find the pairs of fractions that give the answers when subtracted from each other.

$\frac{2}{12}$ $\frac{1}{10}$ $\frac{4}{5}$ $\frac{8}{10}$ $\frac{3}{5}$ $\frac{7}{12}$

a $\frac{7}{10} =$

b $\frac{1}{5} =$

c $\frac{5}{12} =$

Unit 24 **Subtraction of Fractions** (TRB pp. 112–115)
Fractions, decimals and percentages MA3-7NA compares, orders and calculates with fractions, decimals and percentages

Subtraction of Fractions and Number Lines

1 Complete the number line.

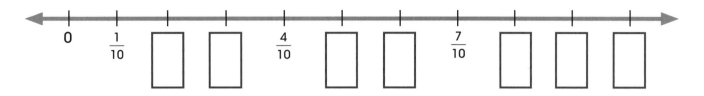

2 Use the number line to answer the equations.

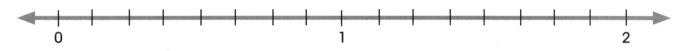

a $\dfrac{7}{8} - \dfrac{2}{8} =$ **b** $\dfrac{6}{8} - \dfrac{5}{8} =$

c $1 - \dfrac{4}{8} =$ **d** $2 - \dfrac{1}{8} =$

e $1\dfrac{5}{8} - \dfrac{6}{8} =$ **f** $1\dfrac{1}{8} - \dfrac{3}{8} =$

3 Use the number line to answer the equations.

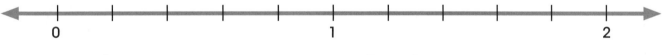

a $1 - \dfrac{3}{5} =$ **b** $\dfrac{7}{5} - \dfrac{2}{5} =$

c $2 - \dfrac{4}{5} =$ **d** $2 - \dfrac{6}{5} =$

4 Complete the equations.

a $\dfrac{9}{10} - \dfrac{4}{10} =$ **b** $2 - \dfrac{7}{8} =$

c $1 - \dfrac{1}{5} =$ **d** $1 - \dfrac{6}{10} =$

e $\dfrac{7}{8} - \dfrac{1}{8} =$ **f** $2 - \dfrac{3}{5} =$

Unit 24 **Subtraction of Fractions** (TRB pp. 112–115)
Fractions, decimals and percentages MA3-7NA compares, orders and calculates with fractions, decimals and percentages

97

Fraction Subtraction Problems

Solve the problems.

1 Grace has $\frac{7}{8}$ of a metre of wire. She cuts $\frac{3}{8}$ of a metre to use in the garden. How much wire does she have left?

2 Arthur has $\frac{3}{4}$ of a cup of flour. He uses $\frac{1}{4}$ of a cup to make a sauce. How much flour does he have left?

3 Emily has $\frac{4}{5}$ of a kilogram of flour and $\frac{2}{5}$ of a kilogram of butter. How much more flour than butter does she have?

4 A recipe requires $\frac{3}{4}$ of a teaspoon of black pepper and $\frac{1}{4}$ of a teaspoon of white pepper. How much more black pepper than white does the recipe need?

5 Alexander runs 1 kilometre on Tuesday and $\frac{1}{3}$ of a kilometre on Wednesday. How much further did he run on Tuesday than on Wednesday?

6 Jill practises soccer for $1\frac{1}{3}$ hours on Friday and $\frac{2}{3}$ of an hour on Saturday. How much more time did she spend practising on Friday than on Saturday?

Extension: How many hours did Jill practise altogether?

 Subtraction of Fractions (TRB pp. 112–115)
Fractions, decimals and percentages MA3-7NA compares, orders and calculates with fractions, decimals and percentages

STUDENT ASSESSMENT

You will need: coloured pencils or felt pens

1 Complete the equations, using the diagrams to help you.

a $\frac{3}{4} - \frac{2}{4} =$ **b** $\frac{7}{10} - \frac{3}{10} =$ **c** $\frac{5}{6} - \frac{4}{6} =$

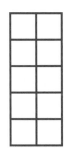
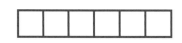

2 Complete the equations.

a $\frac{7}{8} - \frac{5}{8} =$ **b** $\frac{3}{5} - \frac{1}{5} =$ **c** $\frac{7}{9} - \frac{4}{9} =$

d $1 - \frac{2}{4} =$ **e** $1 - \frac{3}{6} =$ **f** $1 - \frac{4}{10} =$

3 Use the number line to solve the equations.

a $\frac{9}{10} - \frac{4}{10} =$ **b** $\frac{6}{10} - \frac{2}{10} =$

c $1 - \frac{9}{10} =$ **d** $\frac{7}{10} - \frac{6}{10} =$

4 William had $\frac{3}{4}$ of a bucket of water. He tipped $\frac{1}{4}$ of a bucket on the lemon tree. How much water did he have left?

Extension: Write your own word problem for the equation $\frac{5}{6} - \frac{3}{6} =$

_____

_____

Unit
24
Subtraction of Fractions (TRB pp. 112–115)
Fractions, decimals and percentages MA3-7NA compares, orders and calculates with fractions, decimals and percentages

99

Time

1 Draw lines to match the pairs of times that are equal.

60 seconds	1 year
31 days	24 hours
1 week	7 days
1 day	1 minute
365 days	1 hour
60 minutes	1 month

2 Order the following times from **shortest** to **longest**:

a 2 minutes, 1 minute, 70 seconds, 90 seconds

_____

b 48 hours, 12 hours, 1 day, 1 week

_____

c a fortnight, 12 days, 1 week, 1 month

_____

3 Circle the **longer** amount of time in each pair.

a 1 minute	120 seconds	**b** 32 hours	2 days
c 1 month	16 days	**d** 95 minutes	1 hour
e 3 weeks	1 month	**f** 60 minutes	1.5 hours

4 Write in the times to complete the table.

	Activity	Time
a	Started school yesterday	
b	Went to bed last night	
c	Had dinner yesterday	
d	Ate lunch yesterday	

Extension: Order the activities in Question 4 from **earliest** to **latest**.

_____

_____

12-Hour Time

1 Complete each clock to show the time.

a 5 o'clock

b half-past 1

c quarter-to 7

d quarter-past 4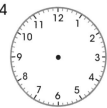

2 Write the digital times.

a

b

c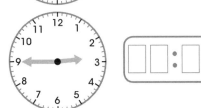

d

3 Write the time shown on each clock.

a  _____

b  _____

c  9:00 _____

d _____

4 Write whether the time shown is **before midday** or **after midday**.

a  am 4:20 _____

b  pm 9:05 _____

c 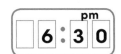 pm 6:30 _____

d  am 11:47 _____

Unit 25 **Time** (TRB pp. 116–119)
Time MA3-13MG uses 24-hour time and am and pm notation in real-life situations, and constructs timelines

101

24-Hour Time

1 Write as **am** or **pm** time.

a 0115 _____

b 1400 _____

c 1730 _____

d 2300 _____

e 0345 _____

f 0615 _____

2 Write the time in 24-hour time.

a 4:45 am _____

b 11:30 pm _____

c 2 pm _____

d 4:00 pm _____

e 11 am _____

f 7:30 am _____

3 Write the morning (**am**) times on the digital clocks in 24-hour time.

a

b

c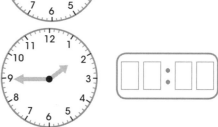

d

4 Write the afternoon/evening (**pm**) times on the digital clocks in 24-hour time.

a

b

c

d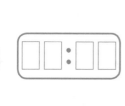

Extension: How much time has passed between the times shown on the clocks in Questions 4a and 4b?

DATE:

STUDENT ASSESSMENT

1 Complete the conversions.

    a 60 seconds = _____ minute

    b 1 day = _____ hours

    c _____ week = 7 days

    d _____ days = 1 fortnight

    e 1 hour = _____ minutes

2 Fill in the missing time on the clocks.

    a

    b

    c

    d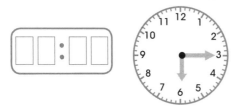

3 Write each time in 24-hour time.

    a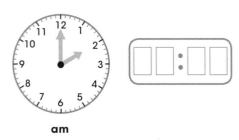

    am

    b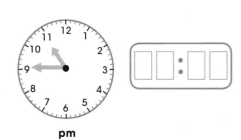

    pm

4 Write as **am** or **pm** time.

    a 0100 _____

    b 1630 _____

    c 2200 _____

    d 0545 _____

Unit
25
Time (TRB pp. 116–119)
Time MA3-13MG uses 24-hour time and am and pm notation in real-life situations, and constructs timelines

103

Time Formats

1 Write as **am** or **pm** time.

    a 0445 _____

    b 1300 _____

    c 2005 _____

    d 1140 _____

    e 0105 _____

    f 1725 _____

2 Convert to 24-hour time.

    a 2:35 am _____

    b 1:10 pm _____

    c 7 pm _____

    d 9:10 pm _____

    e 9:55 am _____

    f 8:25 am _____

3 Write each time in 24-hour time.

    a    pm _____

    b    am _____

    c    am _____

    d    pm _____

4 Find the time that is one hour **later** than the time shown.
Write your answer using 24-hour time.

    a 2:20 am _____

    b 5:40 pm _____

    c 11:55 am _____

    d 9:10 pm _____

Extension: If a train left Melbourne at 11:08 am and arrived
in Ballarat at 12:39 pm, how much time did the trip take?

Using a Stopwatch

You will need: a stopwatch

1 These digital stopwatches show how long it took teams to score their first goal in a soccer tournament.

Team 1 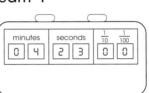   Team 2 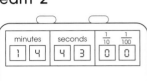   Team 3    Team 4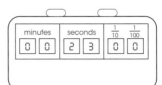

a Which team was the first to score a goal? _____

b Write the longest time in words. _____

c Team 1 scored their second goal 39 seconds after their first. How far into the game was that? _____

2 Sometimes people need to time events to tenths and hundredths of a second, such as in swimming races at the Olympic Games.
Here are some record swimming times. Show each time on a stopwatch.

a 50 metres freestyle:
twenty point one five seconds

b 100 metres freestyle:
forty-eight point nine nine seconds

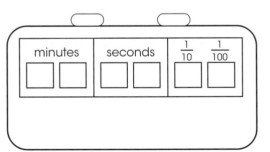   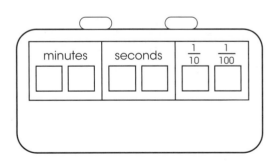

3 Work with a partner for this activity. Time how long it takes for each of you to:

write the alphabet forwards _____

write the alphabet backwards _____

Time Problems

1   Use the schedule of flights from Melbourne to Perth to complete the questions.

Melbourne (depart)	Perth (arrive)
0630	0835
0715	0925
1030	1235
1250	1455
1410	1620

a   Eliza is travelling from Melbourne, and needs to be in Perth by 2 pm.
    What is the **latest** flight that can she take? _____

b   How much time does Eliza's flight take from Melbourne to Perth? _____

c   Jon arrived at Melbourne airport at 9 am.

    i   What is the **earliest** flight that he can take to Perth? _____

    ii What time would he arrive in Perth? _____

2   An express bus runs from the city to Melbourne airport.
    Use the extract from the express bus timetable to complete the questions.

City						
Departure time	0620	0630	0640	0650	0700	0710

a   How often does the bus run? _____

b   If the bus trip takes 20 minutes, and Sue needs to be at the airport at 7:00 am,
    what is the latest bus that she could take? _____

c   If Will takes the 0650 bus to the airport, what is the earliest flight he could take
    to Perth? _____

Extension: Use the timetables above to plan a trip using the express bus and then
a flight from Melbourne to Perth. Record your work on another sheet of paper.

Unit 26 STUDENT ASSESSMENT

1 Find the time that is one hour **later** than the time shown. Write your answer using 24-hour time.

a 4:10 pm _____   **b** 2:30 am _____

c 12:45 pm _____   **d** 11:55 am _____

2 Use the extract of the train timetable from Canberra to Central Sydney to complete the questions.

	A	B	C
Canberra	0643	1155	1703
Queanbeyan	0652	1204	1712
Goulburn	0819	1331	1839
Moss Vale	0910	1422	1930
Campbelltown	1016	1528	2036
Central Sydney	1102	1614	2122

a Alex has to be in Sydney by 4:30 pm. Which is the latest train he can catch from Canberra? _____

b How much time does the trip take between Canberra and Moss Vale? _____

c How much time does the trip take between Goulburn and Central Sydney? _____

d If Ruby catches the train from Goulburn at 6:39 pm, what time will she arrive in Central Sydney? _____

Extension: Write 2 questions based on the train timetable in Question 2.

_____

_____

Unit 26 **Time Problems** (TRB pp. 120–123)
Time MA3-13MG uses 24-hour time and am and pm notation in real-life situations, and constructs timelines

107

Fractions and Decimals

1 Change each fraction to a decimal.

a $\frac{4}{10}$ _____

b $\frac{9}{100}$ _____

c $\frac{50}{100}$ _____

d $\frac{80}{1\,000}$ _____

e $\frac{700}{1\,000}$ _____

f $\frac{18}{100}$ _____

2 Change each decimal to a fraction.

a 0.9 _____

b 0.11 _____

c 0.80 _____

d 0.007 _____

e 0.493 _____

f 0.49 _____

3 Use < or > to complete the number statements.

a $\frac{1}{10}$ ☐ 0.4

b 0.9 ☐ $\frac{9}{100}$

c $\frac{30}{100}$ ☐ 0.8

d 0.12 ☐ $\frac{24}{100}$

e $\frac{8}{1\,000}$ ☐ 0.01

f $\frac{140}{1\,000}$ ☐ 0.4

4 Colour each grid to show the decimal or fraction.

a $\frac{60}{100}$
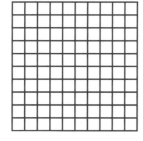

b 0.83
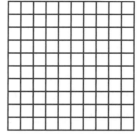

c 0.09
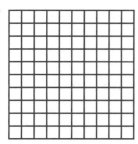

d $\frac{43}{100}$
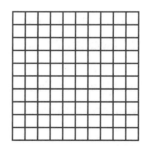

Extension: Order the decimals in Question 2 from **smallest** to **largest**.

Unit 27 **Fractions and Decimals** (TRB pp. 124–127)
Fractions, decimals and percentages MA3-7NA compares, orders and calculates with fractions, decimals and percentages

Decimal and Fraction Patterns – Addition

1 Show the addition number patterns on the number lines.

a Start at 1 and add 0.1 each time.

b Start at 5 and add 0.2 each time.

c Start at 3 and add $\frac{1}{10}$ each time.

d Start at 6 and add $\frac{4}{10}$ each time.

2 Complete the number patterns.

a 3, 3.04, 3.08, 3.12, _____, _____, _____, _____, _____

b 1, 1.3, 1.6, 1.9, 2.2, _____, _____, _____, _____, _____

c $\frac{1}{10}$, $\frac{2}{10}$, $\frac{3}{10}$, $\frac{4}{10}$, _____, _____, _____, _____, _____

d $\frac{44}{10}$, $\frac{46}{10}$, $\frac{48}{10}$, _____, _____, _____, _____, _____

3 Find the rule for each segment of the pattern wheel, and write it in the outer section. One has been done for you.

Extension: On another sheet of paper, complete your own pattern wheel.

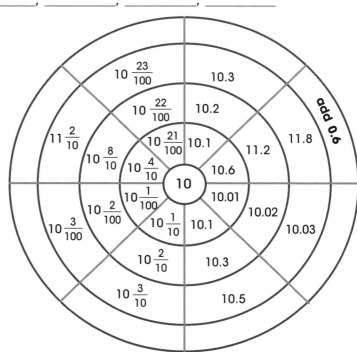

Unit 27 **Fractions and Decimals** (TRB pp. 124–127)
Fractions, decimals and percentages MA3-7NA compares, orders and calculates with fractions, decimals and percentages

109

Decimal and Fraction Patterns – Subtraction

1 Complete the number patterns.

a 1.9, 1.8, 1.7, _____, _____, _____, _____, _____

b 5, 4.8, 4.6, 4.4, _____, _____, _____, _____, _____

c $\frac{50}{100}$, $\frac{45}{100}$, $\frac{40}{100}$, _____, _____, _____, _____, _____

d $1\frac{8}{10}$, $1\frac{7}{10}$, $1\frac{6}{10}$, _____, _____, _____, _____, _____

2 Write a description for each number pattern in Question 1.

a _____

b _____

c _____

d _____

3 Show the subtraction number patterns on the number lines.

a Start at 10 and subtract 0.2 each time.

b Start at 3.2 and subtract 0.4 each time.

c Start at 4 and subtract $\frac{1}{10}$ each time.

d Start at $6\frac{1}{2}$ and subtract $\frac{1}{4}$ each time.

4 Write a subtraction number pattern for:

a decimal subtraction _____

b fraction subtraction _____

Fractions and Decimals (TRB pp. 124–127)
Fractions, decimals and percentages MA3-7NA compares, orders and calculates with fractions, decimals and percentages

STUDENT ASSESSMENT

DATE:

1 Circle the larger amount in each pair.

    a $\frac{1}{2}$ 0.6   **b** 0.15 $\frac{1}{3}$

    c $\frac{4}{10}$ 0.04   **d** 0.9 $\frac{130}{1\,000}$

2 Order each set from **smallest** to **largest** and describe the number pattern.

    a 0.4, 0.2, 0.8, 0.6

    _____ pattern _____

    b 1.6, 2.4, 1.2, 2.0

    _____ pattern _____

    c $\frac{4}{10}$, $\frac{5}{10}$, $\frac{2}{10}$, $\frac{3}{10}$

    _____ pattern _____

    d $\frac{83}{100}$, $\frac{79}{100}$, $\frac{77}{100}$, $\frac{81}{100}$

    _____ pattern _____

3 Write an addition or subtraction number pattern. Record 4 terms.

    a Start at 5 and subtract 0.1 each time.

    _____

    b Start at 5 and add $\frac{2}{10}$ each time.

    _____

    c Start at 1 and subtract $\frac{1}{100}$ each time.

    _____

Unit
27
Fractions and Decimals (TRB pp. 124–127)
Fractions, decimals and percentages MA3-7NA compares, orders and calculates with fractions, decimals and percentages

111

Missing Numbers – Multiplication

1 Record the missing number in each equation.

a $2 \times \boxed{} = 10$ **b** $6 \times 4 = \boxed{}$ **c** $3 \times \boxed{} = 18$

d $\boxed{} \times 10 = 30$ **e** $9 \times \boxed{} = 36$ **f** $5 \times \boxed{} = 40$

g $6 \times \boxed{} = 6$ **h** $\boxed{} \times 8 = 72$ **i** $8 \times 6 = \boxed{}$

2 Record the missing numbers in each equation.

a $2 \times 4 = \boxed{} = 4 \times 2$ **b** $4 \times 3 = \boxed{} = 3 \times \boxed{}$

c $5 \times \boxed{} = \boxed{} = 10 \times 5$ **d** $6 \times 9 = \boxed{} = 9 \times \boxed{}$

e $8 \times \boxed{} = 24 = \boxed{} \times 8$ **f** $\boxed{} \times 5 = 35 = 5 \times \boxed{}$

g $\boxed{} \times 9 = 9 = 9 \times \boxed{}$ **h** $6 \times 7 = \boxed{} = 7 \times \boxed{}$

3 Complete the equations by recording the common answer.

a $3 \times 4 = \boxed{} = 6 \times 2$ **b** $5 \times 2 = \boxed{} = 10 \times 1$

c $6 \times 5 = \boxed{} = 3 \times 10$ **d** $6 \times 4 = \boxed{} = 8 \times 3$

e $3 \times 6 = \boxed{} = 9 \times 2$ **f** $2 \times 3 = \boxed{} = 6 \times 1$

4 Record the missing numbers.

a
$$\begin{array}{r} 3 \\ \times\ \boxed{} \\ \hline 2\,7 \end{array}$$

b
$$\begin{array}{r} \boxed{} \\ \times\ 8 \\ \hline 6\,4 \end{array}$$

c
$$\begin{array}{r} 8 \\ \times\ \boxed{} \\ \hline 4\,8 \end{array}$$

d
$$\begin{array}{r} 4 \\ \times\ 5 \\ \hline \boxed{} \end{array}$$

5 A box contains 9 bags of counters. If there are 90 counters altogether, and each bag contains the same number of counters, how many are in each bag?

Number Sentences (TRB pp. 128–131)
Patterns and algebra MA3-8NA analyses and creates geometric and number patterns, constructs and completes number sentences, and locates points on the Cartesian plane

Missing Numbers – Division

1 Use the arrays to work out the missing numbers.

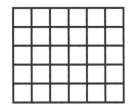   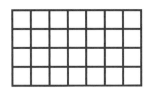   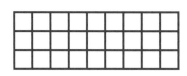

a $5 \times \boxed{} = 30$

$30 = \boxed{} \times 5$

$30 \div \boxed{} = 5$

b $7 \times \boxed{} = \boxed{}$

$\boxed{} = 7 \times \boxed{}$

$\boxed{} \div 7 = \boxed{}$

c $3 \times \boxed{} = \boxed{}$

$\boxed{} = \boxed{} \times 3$

$\boxed{} \div \boxed{} = 3$

2 Record the missing number in each equation.

a $18 \div \boxed{} = 9$

b $80 \div \boxed{} = 10$

c $\boxed{} \div 8 = 7$

d $\boxed{} \div 9 = 8$

e $15 \div 3 = \boxed{}$

f $32 \div \boxed{} = 4$

g $36 \div \boxed{} = 6$

h $21 \div \boxed{} = 3$

i $3 \div \boxed{} = 1$

3 Complete the equations by working out the common answer.

a $21 \div 7 = \boxed{} = 9 \div 3$

b $80 \div 8 = \boxed{} = 10 \div 1$

c $50 \div 10 = \boxed{} = 25 \div 5$

d $56 \div 7 = \boxed{} = 16 \div 2$

4 Record the missing numbers.

a $\boxed{} \overline{)\,28\,}^{\,7}$

b $9\,\overline{)\,63\,}^{\,\boxed{}}$

c $6\,\overline{)\,\boxed{}\,}^{\,5}$

5 One hundred children are going on an excursion. If they are divided into 10 even groups, how many children are in each group?

_____

6 Fifty-four eggs are to be packed into cartons of 6.
How many cartons are needed?

_____

Unit 28 **Number Sentences** (TRB pp. 128–131)
Patterns and algebra MA3-8NA analyses and creates geometric and number patterns, constructs and completes number sentences, and locates points on the Cartesian plane

113

Solving Equations

DATE:

1 Write a number sentence for each problem. Then solve the number sentence to find the missing number.

    a Multiply *a number* by 8 to get 80.

    b Divide *a number* by 3 to get 9.

    c Multiply *a number* by 10 to get 1.

    d Divide *a number* by 6 to get 6.

    e Add 3 to *a number* and then multiply by 4 to get 24.

    f Subtract 5 from a *number* and then multiply by 10 to get 40.

2 Write a number problem for each number sentence.

    a $9 \times 6 =$ _____

    b $40 \div 8 =$ _____

Extension: Write a number sentence for each problem. Use a calculator to solve the number sentence.

    1 16 multiplied by the difference of 9.6 and 7.3

    2 21 divided by the sum of 1.3 and 2.9

Number Sentences (TRB pp. 128–131)
Patterns and algebra MA3-8NA analyses and creates geometric and number patterns, constructs and completes number sentences, and locates points on the Cartesian plane

Unit 28 STUDENT ASSESSMENT

1 Record the missing number in each equation.

a 10 × ☐ = 50

b 9 × ☐ = 63

c ☐ × 8 = 40

d ☐ × 4 = 32

e 35 ÷ ☐ = 5

f 10 ÷ ☐ = 1

g ☐ ÷ 6 = 3

h ☐ ÷ 7 = 4

2 Record the missing numbers.

a
```
      6
  × ☐
  ────
   2 4
```

b
```
    ☐
  ×   5
  ────
   2 5
```

c
```
      6
  × 7
  ────
   ☐
```

d
```
        10
  ☐ ) 100
```

e
```
      ☐
  6 ) 4 8
```

f
```
        9
  4 ) ☐
```

3 Draw a line to match the equations with the same answer.

a 2 × 1   81 ÷ 9

b 2 × 4   8 ÷ 4

c 5 × 2   64 ÷ 8

d 3 × 3   60 ÷ 6

4 Write a number sentence for each problem. Then solve the number sentence to find the missing number.

a Multiply *a number* by 3 to get 24.

b Multiply *a number* by 7 and then add 4 to get 18.

c Divide *a number* by 8 to get 7.

d Divide *a number* by 3 and then subtract 1 to get 5.

Unit 28 **Number Sentences** (TRB pp. 128–131)
Patterns and algebra MA3-8NA analyses and creates geometric and number patterns, constructs and completes number sentences, and locates points on the Cartesian plane

115

Vegetable Plant Audit

The table below shows data about a school vegetable garden.

1 Complete the blank cells in the table.

Vegetable plant	Number in garden bed 1	Number in garden bed 2	Total number of plants
Carrot	20	25	
Bean	10	15	
Broccoli	20	10	
Cauliflower	15	15	
Corn	0	20	
Total			

2 Represent the data as a column graph. Don't forget to add the labels!

3 a Which vegetable plant did the school have **most** of? _____

b Which vegetable plant did the school have **least** of? _____

4 What vegetable plant would you recommend be added to the vegetable garden? _____ Explain why. _____

_____

Data Display (TRB pp. 132–135)
Data MA3-18SP uses appropriate methods to collect data and constructs, interprets and evaluates data displays, including dot plots, line graphs and two-way tables

Presenting Data

You will need: a ruler, your researched data

1 Name the location (i.e. the website) where you found your data. _____

2 Present your data as a table.

3 Draw an appropriate graph to represent your data.

4 Write 3 statements about your data.

a _____

b _____

c _____

Unit
29
Data Display (TRB pp. 132–135)
Data MA3-18SP uses appropriate methods to collect data and constructs, interprets and evaluates data displays, including dot plots, line graphs and two-way tables

117

Organising Data

1 My group's question is:

_____

2 Here is our collected data.

3 We have decided to present our data as
(e.g. table, line graph, column graph): _____

4 The hardest part of the task was: _____

_____

5 The part of the task I enjoyed most was: _____

_____

6 As a group, we worked: _____

_____

7 When I present, I will be speaking about: _____

_____

Data Display (TRB pp. 132–135)
Data MA3-18SP uses appropriate methods to collect data and constructs, interprets and evaluates data displays,
including dot plots, line graphs and two-way tables

STUDENT ASSESSMENT

You will need: a ruler

Look at the data about sports played by a group of children.

Name	Sport
Jack	soccer, football
Hayley	football, netball
Sarah	soccer, cricket
Claire	football
Henrick	cricket

Name	Sport
Georgia	soccer, netball
Ahmed	football
Tyler	soccer
Jordan	cricket
Cameron	cricket

Name	Sport
Ling	netball
Steve	soccer, football
Rasdeep	netball
Rhys	soccer, cricket
Rachel	netball

1 What question(s) could have been asked to collect the data?

_____

2 Organise the data into a tally table.

3 Draw a graph to represent the data.

4 Write 3 questions about the graph.

a _____

b _____

c _____

Unit **29** **Data Display** (TRB pp. 132–135)
Data MA3-18SP uses appropriate methods to collect data and constructs, interprets and evaluates data displays, including dot plots, line graphs and two-way tables

119

School Canteen Data

You will need: a ruler

1 My group's members are: _____

_____

2 My group's question/topic is: _____

3 Here is our collected data.

4 Here is a graph of our data.

5 Three things that we discovered from our work are:

a _____

b _____

c _____

Interpreting Data (TRB pp. 136–139)
Data MA3-18SP uses appropriate methods to collect data and constructs, interprets and evaluates data displays, including dot plots, line graphs and two-way tables

School Water Use

1 My partner's name is: _____

2 Our investigation/topic is: _____

3 Here is our collected data.

4 Here is how we organised our data.

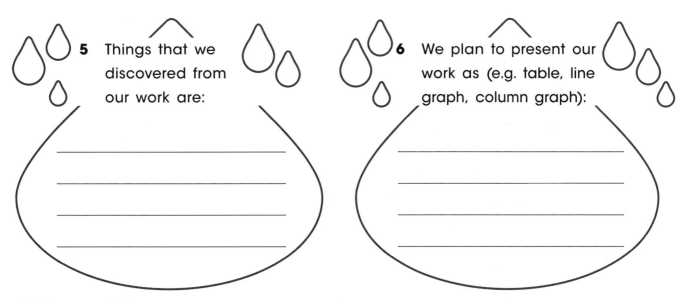

5 Things that we discovered from our work are:

_____
_____
_____
_____

6 We plan to present our work as (e.g. table, line graph, column graph):

_____
_____
_____
_____

Unit 30 **Interpreting Data** (TRB pp. 136–139)
Data MA3-18SP uses appropriate methods to collect data and constructs, interprets and evaluates data displays, including dot plots, line graphs and two-way tables

121

Is it Fair?

The following computer log was kept by Year 5KG over 1 week.

Name	Logged on	Logged off	Duration of session	Name	Logged on	Logged off	Duration of session
Luke	9:05	9:45		Luke	11:00	12:00	
Tara	9:45	10:30		Ahmed	12:00	12:30	
Ali	11:00	11:20		Fiona	1:30	2:30	
Luke	11:20	12:30		Hannah	2:30	3:00	
Ryan	1:30	2:30		Ryan	9:00	9:45	
Ahmed	2:30	3:00		Hannah	9:45	10:30	
Jodie	3:00	3:30		Luke	11:00	11:15	
Tara	9:15	10:00		Tara	11:15	12:00	
Ben	10:00	10:30		Tara	1:30	3:30	

Use the data to determine whether the computer time was shared fairly in the class. You could draw a graph or create a table below to support your point of view.

Explain why you presented your point of view as you did.

_____

_____

Interpreting Data (TRB pp. 136–139)
Data MA3-18SP uses appropriate methods to collect data and constructs, interprets and evaluates data displays, including dot plots, line graphs and two-way tables

STUDENT ASSESSMENT

The following table presents data about coloured donation tokens that were collected over 5 days.

Each token had a value: red = $1, blue = $2, green = $5 and black = $0.50

	Red	Blue	Green	Black
Monday	18	27	6	40
Tuesday	21	30	10	25
Wednesday	16	32	21	45
Thursday	40	35	19	35
Friday	19	20	4	20

1 a Which colour raised the most money? _____

   b On which day was the most money raised? _____

2 Create a graph or list to justify your responses to Question 1.

Unit
30
Interpreting Data (TRB pp. 136–139)
Data MA3-18SP uses appropriate methods to collect data and constructs, interprets and evaluates data displays, including dot plots, line graphs and two-way tables

123

Receipts

Use the receipts from two different stores to answer the questions.

Stationery Store

Store 1136

Tax Invoice

Student chair	$39.67
Pens (pk 6)	$4.95
Photocopy paper Quantity: 2 @ $6.75	$13.50
Total	$58.12
Payment by Cash	$60.00
Change	$1.88
	($1.90)
GST included in total	$5.81

Cool Dude Clothes

Store 1103

Tax Invoice

Hat	$16.95
T-shirt (L)	$32.65
Jeans (L)	$54.87
Subtotal	$104.47
Discount 5%	
Total savings	−$5.22
Total	$99.25
(includes $9.93 GST)	
Payment	$100.00
Change	$0.75
Please retain receipt for returns.	

1 What does the T-shirt cost? _____

2 In which store was the chair bought? _____

3 How many packets of paper were bought? _____

4 What size pants were purchased? _____

5 What was the total of the 2 receipts? _____

6 What payment method was used for the items? _____

Extension: Write 3 questions of your own about the receipts.

_____

_____

_____

Fundraising Budget

DATE:

1 My class' activity is to: _____

_____

2 My group's task is to investigate: _____

_____

3 Things that we discovered are: _____

_____

4 Here is my class' fundraising budget.

5 Other areas we may need to investigate are: _____

_____

Unit 31
Financial Plans (TRB pp. 140–143)
Addition and subtraction (money) MA3-5NA selects and applies appropriate strategies for addition and subtraction with counting numbers of any size

125

GST

You will need: a calculator

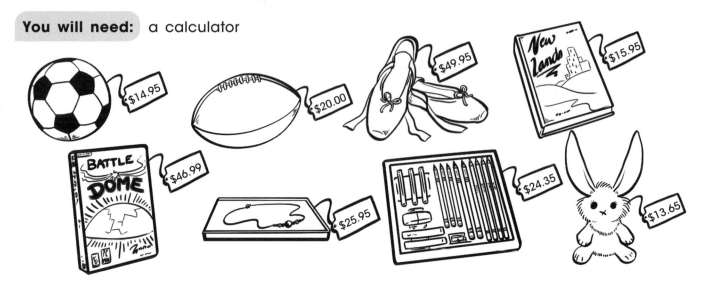

1 Of the items shown, select 2 that you would like to buy.

_____

2 **a** Add your chosen items to the receipt below. _____

   b Complete the change information and the GST details.
      Use the working space provided.

Things for You!

Tax Invoice

_____ $ _____

_____ $ _____

Total $ _____

Payment $100.00

Change $ _____

GST inc. in total $ _____

Working space

Financial Plans (TRB pp. 140–143)
Addition and subtraction (money) MA3-5NA selects and applies appropriate strategies for addition and subtraction
with counting numbers of any size

STUDENT ASSESSMENT

DATE:

A football club wants to hold a fundraiser.

1 What type of fundraiser could it be? _____

_____

2 a List 5 things that could be purchased for the fundraiser, and add them to the table.

   b Estimate the cost of each of the 5 items, and include this on the table.

	Items to purchase	Estimated cost ($)
1		
2		
3		
4		
5		
	Total cost =	

3 Work out the total cost of the 5 items, and include it on the table.

4 What other things may need to be considered for the budget?

_____

_____

5 If 5 football jumpers were purchased for the fundraiser at $45 per jumper, work out the following:

   a the total cost of the jumpers. _____

   b the total GST paid on the jumpers. _____

6 Do you think the fundraiser could be successful? Explain your answer.

_____

_____

Unit
31
Financial Plans (TRB pp. 140–143)
Addition and subtraction (money) MA3-5NA selects and applies appropriate strategies for addition and subtraction with counting numbers of any size

127

Symmetry

1 Complete the table.

	Shape	Number of sides	Number of lines of symmetry
a			
b			
c	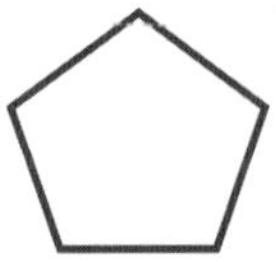		
d	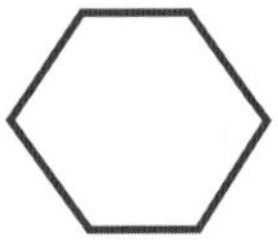		
e			
f	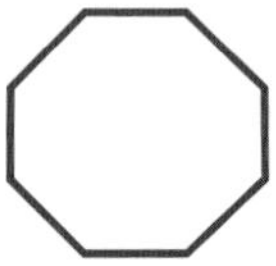		
g	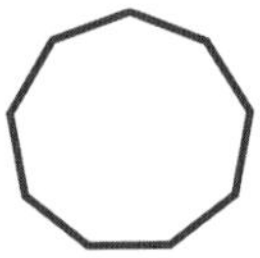		

2 Do you see a pattern between the number of sides of a shape and the number of lines of symmetry it has? _____ Explain your findings.

_____

_____

3 How many lines of symmetry do you think a 20-sided shape would have? _____ Justify your answer.

_____

_____

Transformations (TRB pp. 144–147)
Two-dimensional space MA3-15MG manipulates, classifies and draws two-dimensional shapes, including equilateral, isosceles and scalene triangles, and describes their properties

Reflections and Translations

1   Imagine that the dotted line is a mirror. Draw the reflection of each shape on the other side of the line.

a    b    c

d 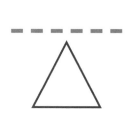   e    f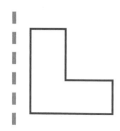

2   Translate each shape in the direction indicated by the arrow.

a 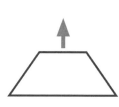   b 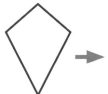   c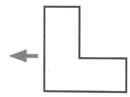

d    e 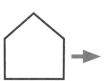   f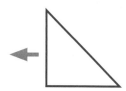

3   Write your name 2 different ways.

   a reflected (down)   b translated to the right

Rotations

1 Complete the table by drawing each shape rotated by the stated amount.

	Shape	Rotated by 90°	Rotated by 180°
a	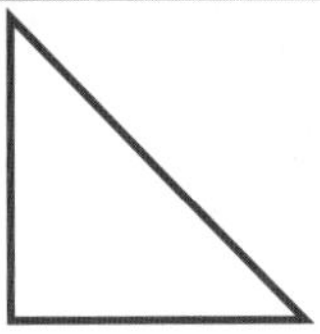		
b	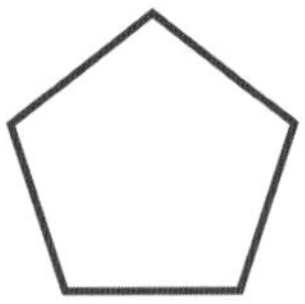		
c			
d	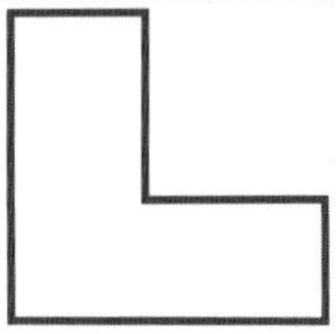		
e			
f			
g	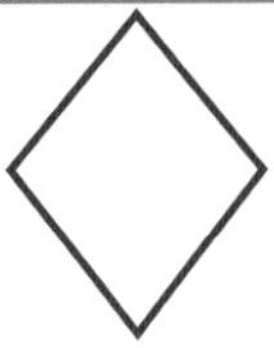		

2 List 5 **lower case** letters that read exactly the same when rotated 180°.

_____

3 List 5 **upper case** letters that read exactly the same when rotated 180°.

_____

Transformations (TRB pp. 144–147)
Two-dimensional space MA3-15MG manipulates, classifies and draws two-dimensional shapes, including equilateral, isosceles and scalene triangles, and describes their properties

STUDENT ASSESSMENT

1 Draw the line(s) of symmetry for each shape.

a   **b**  **c**

d  **e**  **f**

2 Identify whether a reflection or translation is shown.

a 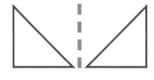 **b** 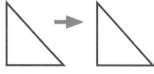 **c**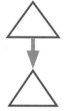

_____ _____ _____

d 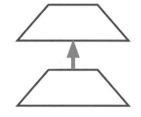 **e**  **f**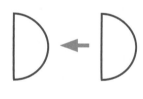

_____ _____ _____

3 Rotate each shape as indicated.

a to the left   **b** to the right

c 180°  **d** 90°

Extension: List 5 letters of the alphabet that are symmetrical.

_____

Unit
32
Transformations (TRB pp. 144–147)
Two-dimensional space MA3-15MG manipulates, classifies and draws two-dimensional shapes, including equilateral,
isosceles and scalene triangles, and describes their properties

131

Tessellations

You will need: coloured pencils or felt pens

Investigate to see if each shape will tessellate. Use different colours to find out.

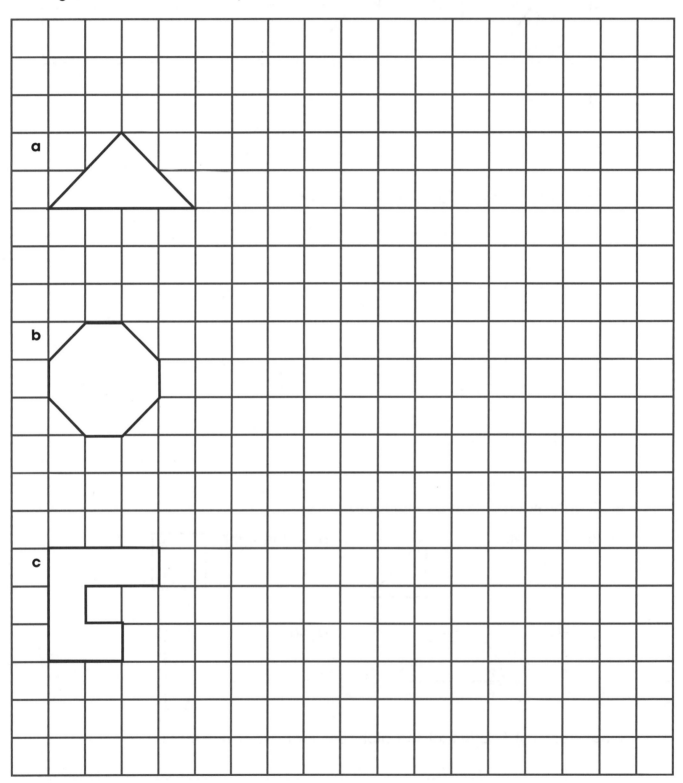

Use of Transformations (TRB pp. 148–151)
Two-dimensional space MA3-15MG manipulates, classifies and draws two-dimensional shapes, including equilateral, isosceles and scalene triangles, and describes their properties

Transforming Shapes

DATE:

1 List the movements that were made to tessellate shape A to shape F.
The first step has been done for you in Question a.

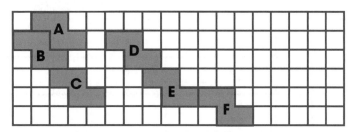

a A to B: translated one space to the left and one space down.

B to C: _____

C to D: _____

D to E: _____

E to F: _____

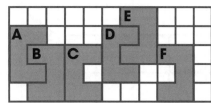

b A to B: _____

B to C: _____

C to D: _____

D to E: _____

E to F: _____

2 Create your own transformation that includes **at least** one rotation, and label each movement from A to F.

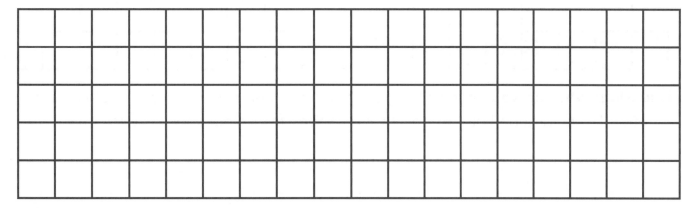

Unit **33**

Use of Transformations (TRB pp. 148–151)
Two-dimensional space MA3-15MG manipulates, classifies and draws two-dimensional shapes, including equilateral, isosceles and scalene triangles, and describes their properties

133

Enlarging 2D Shapes

Enlarge each shape by the factor indicated. For example, as the square has 2 written on it, all of the square's dimensions will be doubled.

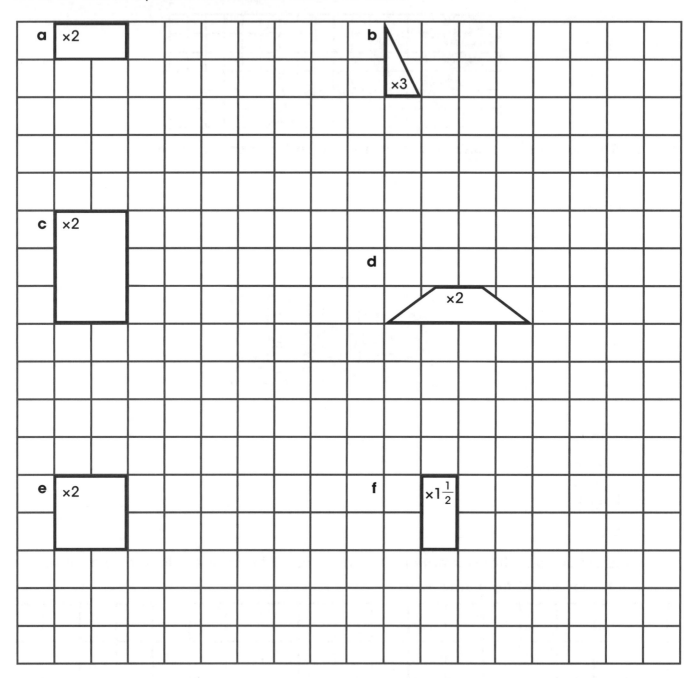

Extension: What do you notice about each shape's perimeter and area before and after it is enlarged? Refer to one of the shapes to explain your answer.

_____

_____

Unit 33 **Use of Transformations** (TRB pp. 148–151)
Two-dimensional space MA3-15MG manipulates, classifies and draws two-dimensional shapes, including equilateral, isosceles and scalene triangles, and describes their properties

STUDENT ASSESSMENT

1 Explain what each word means.

   a tessellation _____

   b enlargement _____

   c rotation _____

2 Draw an example of each of the following:

Tessellation					Enlargement					Rotation				

3 Complete the tasks using this triangle:

   a rotate it to the right

   b check to see if it tessellates   **c** enlarge it by 2

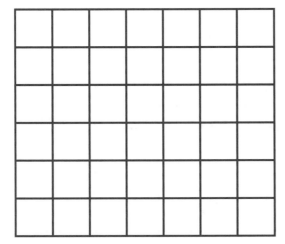 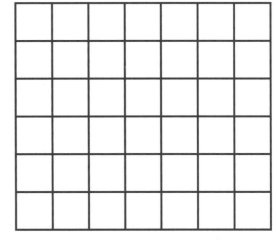

Use of Transformations (TRB pp. 148–151)
Two-dimensional space MA3-15MG manipulates, classifies and draws two-dimensional shapes, including equilateral, isosceles and scalene triangles, and describes their properties

135

Challenges

Magic Squares

Complete the magic squares so that each row and column totals the number indicated below the square.

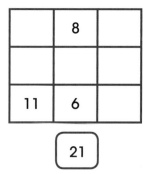

21

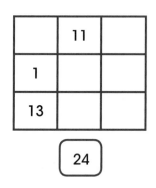

24

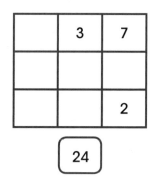

24

Number Trail

Solve the number trail.

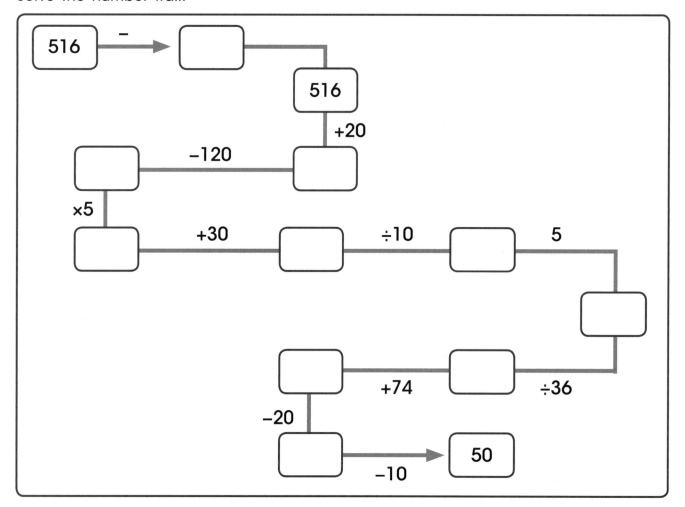

Challenges

Finding the Number Combination

Fill in the squares with only even numbers between 2 and 10.
No row, column or diagonal can use the same number more than once.

2	4		8	10
				6
	8			

Calculator

Can you use only 4 buttons to produce this number on the calculator?

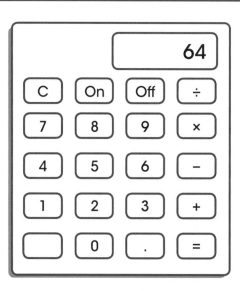

Digits 1–4

Use the digits 1, 2, 3 and 4 and any operation to find all of the numbers from 1 to 20. You can combine the digits but not use them more than once for each question.

For example: (4 – 2) + (3 – 1) = 4

24 ÷ (3 + 1) = 6

Challenges

Quick Quizzes

Quiz A	Quiz B
1 7, 14, ☐, ☐, 35	**1** 6, 12, 18, ☐, ☐, 36
2 7 × 8 =	**2** 9 × 6 =
3 9 × 4 =	**3** 4 × 7 =
4 20 ÷ 5 =	**4** 56 ÷ 7 =
5 The area of a square with side length 5 cm is _____	**5** The area of a rectangle with side lengths 4 cm and 8 cm is _____
6 (5 + 7) × 3 =	**6** (9 × 2) – (3 × 1) =
7 6 × (3 + 1) =	**7** 4 × (5 + 3) =
8 50% of 60 =	**8** 25% of 40 =
9 9.2 – 3.1 =	**9** 7.3 – 5.2 =
10 4.5 + 2.3 =	**10** 5.6 + 3.2 =
11 $4.50 + $3.25 =	**11** $5.35 + $2.60 =
12 6.5 m = ☐ cm	**12** ☐ m = 250 cm
13 subtract 25 from 200	**13** divide 120 by 60 =
14 product of 8 and 3 =	**14** the difference between 25 and 57 =
15 A quarter of an hour after 2.10 pm is:	**15** half-past 7 written in digital format is:

Making a Half

How many different ways can you represent a half, using the square below?
(Hint: There is more than 10!)

Challenges

Hexagons and Perimeters

This regular hexagon has a perimeter of 24 cm. The sides of the hexagon are joined together, what could the perimeter of this new shape be?

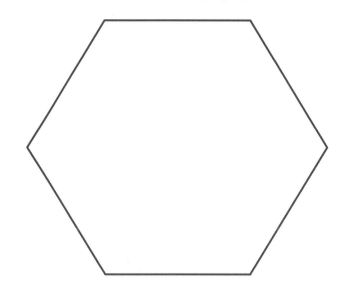

Changing Perimeter

Does the perimeter of the shape change if you cut out a rectangle from the shape?

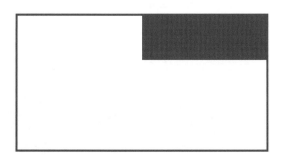

Re-arranging Numbers

Can you rearrange the 16 numbers so that the total of the four 4-digit numbers is as close to 9000 as possible?

	1	2	3	4
	1	2	3	4
	1	2	3	4
+	1	2	3	4
=	4	9	3	6

Challenges

How Many Squares? How Many Rectangles?

Here is a block of 16 small squares.
There are other squares in the shape,
such as the shaded one.

- How many squares can you find?

- How many rectangles can you find?

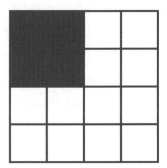

Jars and Pencils

I have some pencils and some jars.

If I put 4 pencils in each jar, I will have 1 jar left over.
If I put 3 pencils in each jar, I will have one 1 left over.
How many pencils and how many jars do I have?

What Coins Do I Hold?

What are the possible combinations of having
3 coins in your hand? List all of the possibilities.

Filling the Gaps

Fill in the empty balls.
The numbers in the two balls below add to give the one above.

a
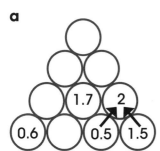

b
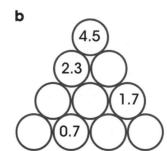

Challenges

Finding the Weight

You have 12 balls identical in size and appearance but one is an odd weight
(it could be either light or heavy). You have a set of scales (balance) which will
give three possible readings:

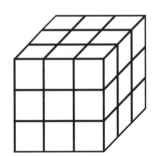

- Left = Right

- Left > Right

- Left < Right

(Left and Right have equal weight, Left is Heavier, or Left is Lighter.).
You have only three chances to weigh the balls in any combination using the scales.
Determine which ball is the odd one and if it's heavier or lighter than the rest.
Explain how you would do it.

Cutting a 3 × 3 × 3 Cube

Imagine a 3 × 3 × 3 cube.
How many cuts do you need to make
to break it into 27 1 × 1 × 1 cubes?
(A cut may go through multiple pieces.)

Ten Coins in Five Rows Puzzle

The arrangement below shows 10 coins in three rows of four. Your task is to re-arrange
the same ten coins to make five rows of four. Each line must be straight and must
contain exactly four coins.

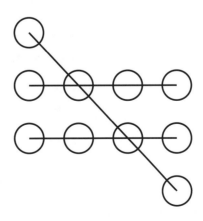

Maths Glossary

acute angle an angle less than 90°

analogue clock a clock with two hands and 12 numerals

angle the space between the intersection of two straight lines

area the size of a surface; calculated with the formula: length × width

array a set of items arranged in rows and columns

axes the lines that form the framework of a graph, i.e. x-axis and y-axis

bar graph (column graph) a graph that has horizontal or vertical bars to represent the data

BODMAS gives the order for solving equations; it stands for: **B**rackets, **O**f, **D**ivision, **M**ultiplication, **A**ddition, **S**ubtraction

capacity how much a container can hold

chance the likelihood something will happen

composite shape a shape consisting of a number of regular shapes

cone a solid with a circular base that comes to a point at the top

coordinates a pair of numbers and/or letters that represents a position on a map or graph

cube a solid with six equal square faces and eight corners

cylinder a solid with two circular faces at right angles to a curved surface

data factual information gathered for research

decimal having ten parts

decimal fraction a fraction written as a decimal, e.g. $\frac{1}{10}$ = 0.1

degrees the unit for measuring an angle in geometry

denominator the bottom number in a fraction, which shows how many parts make up a whole

difference the amount by which two numbers differ; the answer to a subtraction problem

dividend a number that is divided by another number

divisor a number that is divided into another number

equivalent fractions fractions that represent the same amount, e.g. $\frac{1}{2} = \frac{2}{4} = \frac{3}{6} = \frac{4}{8}$

Maths Glossary

estimate a guess based on past experience

face a flat surface on a three-dimensional shape

factor a whole number that can be divided exactly into another number, e.g. the factors of 8 are 1, 2, 4 and 8

fraction a part of a whole quantity or number

horizontal line a line that runs parallel to the horizon

improper fraction a fraction with a numerator greater than the denominator

inverse operation the opposite operation, i.e. subtraction and addition *or* multiplication and division

irregular shape a shape that is not regular

line graph a graph drawn as points joined to form a line

line of symmetry a line drawn across the centre of a shape so that each half of the shape is the mirror image of the other

mass quantity of matter in an object

mixed number a number that consists of a whole number and a fraction

multiple the result of multiplying a number by an integer (not a fraction), e.g. 12 is a multiple of 3

negative number a number less than zero; negative numbers are preceded by a minus sign, e.g. –4

net a flat shape that can be folded into a 3D object

number line a line marked with numbers to show operations or patterns

number sentence a mathematical sentence written with numbers and mathematical symbols

number sequences an ordered set of numbers

numerator the top number in a fraction, which shows how many parts of the whole there are

obtuse angle an angle greater than 90°

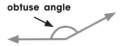

parallel lines a set of lines that remain the same distance apart and do not intersect

perimeter the distance around the edge of a 2D shape

perpendicular line a vertical line that makes a right angle where it meets a horizontal line

pictograph a graph in which data is represented by pictures

pie graph (sector graph) a graph in which a circle is divided into sections ('pieces of pie') to represent data

Maths Glossary

place value the value of a digit dependent on its position in a number

positive number a number greater than zero

prime number a whole number that can be divided only by itself and 1

prism a 3D object with two parallel ends of the same size and shape, e.g. rectangular prism, triangular prism

probability the chance a particular outcome will occur compared to all outcomes

product the answer to a multiplication problem

pyramid a 3D object in which the base is a polygon and all other faces are triangles

quotient the answer to a division problem

reflex angle an angle between 180° and 360°

regular shape a shape in which all sides are equal and all angles are equal

remainder the amount left over after one number has been divided into another

right angle an angle of exactly 90°

rotation turning an object around a fixed point

scale the ratio in which something is represented as either greater or smaller than lifesize

simple fraction a fraction such as $\frac{1}{2}, \frac{1}{3}, \frac{3}{7}$; also called a 'common fraction' or 'vulgar fraction'

sphere a solid shaped like a ball

symmetry a shape has symmetry if both its parts match when folded along a line

table (data) information organised in columns and rows

temperature a measurement of how hot or cold something is

transformation a change in position or size including: translation, rotation, reflection or enlargement (zoom)

translation the movement of a shape that occurs without flipping or reflecting it

unit fraction a fraction with a numerator of 1, e.g. $\frac{1}{2}$ or $\frac{1}{10}$

Venn diagram overlapping circles used to show different sets of information

vertex the point at which two or more lines meet to form an angle or corner

vertical line a line that runs at right angles to the horizon

volume the space occupied by a 3D object

weight mass affected by gravity